HOW TO STUDY AS A
MATHEMATICS MAJOR

HOW TO STUDY AS A
MATHEMATICS MAJOR

LARA ALCOCK

Mathematics Education Centre, Loughborough University, UK

OXFORD
UNIVERSITY PRESS

OXFORD
UNIVERSITY PRESS

Great Clarendon Street, Oxford, OX2 6DP,
United Kingdom

Oxford University Press is a department of the University of Oxford.
It furthers the University's objective of excellence in research, scholarship,
and education by publishing worldwide. Oxford is a registered trade mark of
Oxford University Press in the UK and in certain other countries

First Edition published in 2013

Impression: 3

British Library Cataloguing in Publication Data

Data available

ISBN 978-0-19-966131-2

Printed in Great Britain by
Clays Ltd, St Ives plc

PREFACE

Every year, thousands of students declare mathematics as their major. Many of these students are extremely intelligent and hard-working. However, even the best struggle with the demands of making the transition to advanced mathematics. Some struggles are down to the demands of increasingly independent study. Others, however, are more fundamental: the mathematics shifts in focus from calculation to proof, and students are thus expected to interact with it in different ways. These changes need not be mysterious—mathematics education research has revealed many insights into the adjustments that are necessary—but they are not obvious and they do need explaining.

This book aims to offer such explanation for a student audience, and it differs from those already aimed at similar audiences. It is not a popular mathematics book; it is less focused on mathematical curiosities or applications, and more focused on how to engage with academic content. It is not a generic study skills guide; it is focused on the challenges of coping with formal, abstract undergraduate mathematics. Most importantly, it is not a textbook. Many "transition" or "bridging" or "foundations" textbooks exist already and, while these do a good job of introducing new mathematical content and providing exercises for the reader, my view is that they still assume too much knowledge regarding the workings and values of abstract mathematics; a student who expects mathematics to come in the form of procedures to copy will not know how to interact with material presented via definitions, theorems, and proofs. Indeed, research shows that such a student will likely ignore much of the explanatory text and focus disproportionately on the obviously symbolic parts and the exercises. This book aims to head off such problems by starting where the student is; it acknowledges existing skills, points out common experiences and expectations, and re-orients students so that they know what to look for in texts and lectures on abstract mathematics. It could thus be considered a universal prelude to upper-level textbooks in general and to standard transition textbooks in particular.

Because this book is aimed at students, it is written in the style of a friendly, readable (though challenging and thought-provoking) self-help book. This means that mathematicians and other mathematics teachers will find the style considerably more narrative and conversational than is usual in mathematics books. In particular, they might find that some technical details that they would emphasize are glossed over when concepts are first introduced. I made a deliberate decision to take this approach, in order to avoid getting bogged down in detail at an early stage and to keep the focus on the large-scale changes that are needed for successful interpretation of upper-level mathematics. Technical matters such as precise specification of set membership, of function domains, and so on, are pointed out in footnotes and/or separated out for detailed discussion in the later chapters of Part 1.

To lead students to further consideration of such points, and to avoid replicating material that is laid down well elsewhere, I have included a further reading section at the end of each chapter. These lists of readings aim to be directive rather than exhaustive, and I hope that any student who is interested in mathematics will read widely from such material and thus benefit from the insights offered by a variety of experts.

This book would not have been possible without the investigations reported by the many researchers whose works appear in the references. My sincere thanks also to Keith Mansfield, Clare Charles, and Viki Mortimer at Oxford University Press, to the reviewers of the original book proposal, and to the following colleagues, friends, and students who were kind enough to give detailed and thoughtful feedback on earlier versions of this work: Nina Attridge, Thomas Bartsch, Gavin Brown, Lucy Cragg, Anthony Croft, Ant Edwards, Rob Howe, Matthew Inglis, Ian Jones, Anthony Kay, Nathalie Matthews, David Sirl, and Jack Tabeart. Thanks in particular to Matthew, who knew that I was intending to write and who gave me related books for my birthday in a successful attempt to get me started.

Finally, this book is dedicated to my teacher George Sutcliff, who allowed me to find out how well I could think.

CONTENTS

Part 2 Study Skills

SYMBOLS

Symbol	Meaning	Section
$\mathbb{N}$	the set of all natural numbers	2.3
$\mathbb{Z}$	the set of all integers	2.3
$\mathbb{Q}$	the set of all rational numbers	2.3
$\mathbb{R}$	the set of all real numbers	2.3
$\mathbb{C}$	the set of all complex numbers	2.3
$\in$	is an element of	2.3
$\subseteq$	is a subset of	2.3
ϕ	phi (Greek letter often used for a transformation)	2.4
$\mathbb{R}^4$	the set of all 4-component vectors	2.4
$f : \mathbb{R} \to \mathbb{R}$	function f from $\mathbb{R}$ to $\mathbb{R}$	3.6
$\phi : U \to V$	transformation ϕ from the set U to the set V	4.4
$\Rightarrow$	implies	4.5
$\Leftrightarrow$	is equivalent to (or "if and only if")	4.5
$\forall$	for all	4.7
$\exists$	there exists	4.7
$\notin$	is not an element of	6.3
Σ	sigma (Greek letter used for a sum)	6.4
$\emptyset$	the empty set	8.4
$[a, b]$	closed interval	8.5
(a, b)	open interval	8.5
$\{a, b, c\}$	set containing the elements a, b and c	8.5
$\{x \in \mathbb{R} \mid x^2 < 2\}$	set of all real x such that $x^2 < 2$	8.5

INTRODUCTION

This short introduction explains the aim and structure of this book, and suggests that different groups of readers might like to approach the chapters in different orders. For those who have not yet begun their undergraduate studies, it also explains some useful vocabulary.

AIM OF THE BOOK

This book is about how to make the most of a mathematics major. It is about the nature of undergraduate mathematics, about the ways in which professors expect students to think about it, and about how to keep on top of studying while enjoying undergraduate life. It is written for those who intend to study for a mathematics major, and for those who have already started.

If you are among the first group, you are probably in one of two positions. You might be a bit nervous about the whole business. Perhaps you have done well in mathematics so far, but you think that your success is mostly down to hard work. Perhaps you believe that others have some innate mathematical talent that you lack, and that in upper-level courses you will be in classes full of geniuses and will end up being found out as a fraud. As a mathematics professor, I meet quite a lot of students like this. Some of them always doubt themselves, and they get their degrees but they don't really enjoy their studies. Others do come to realize that their thinking is as good as that of anyone else. They develop more faith in themselves, they succeed, and they enjoy the whole process of learning. If you are a bit nervous, I hope that this book will help you to feel prepared, to make good progress, and to end up in this latter group.

You might, on the other hand, be confident that you are going to succeed. That's how I felt when I began my undergraduate studies in the UK. I'd always been the best student in my mathematics classes, I had no problems with the extra mathematics I took in high school, and I was pretty sure I wanted to be a mathematician. But when I arrived at college and began taking the equivalent of upper-level mathematics courses, I was forced to adjust my expectations. For a while I thought that I would only barely merit a degree and that I should seriously downgrade my career aspirations. Then, in a very gratifying turnaround, I got the hang of advanced mathematics and was eventually awarded what is known as a "first class" degree. This was largely due to a few key insights that I gained from some excellent teachers. In fact, these insights prompted me to decide that studying how people think about mathematics is even more interesting than studying mathematics, so I went on to do a PhD in Mathematics Education. These days I give lectures on undergraduate mathematics, and spend the rest of my time conducting research studies to investigate how people learn and think about it.

One simple but important thing I have learned is that, whatever their feelings on declaring mathematics as their major, most students have a lot to learn about how to study it effectively. Even those who end up doing very well are usually somewhat inefficient to start with. That's why I'm writing this book: to give you a leg-up so that your academic life is easier and more enjoyable than it would otherwise be.

However, this book is not about some magical easy way to complete a mathematics major without really trying. On the contrary, a lot of hard work will be required. But this is something to embrace. A mathematics major *should* be challenging—if it were easy, everyone would have one. And, if you've got this far in your studies, you must have experienced the satisfaction of mastering something that you initially found difficult. The book is, however, about how to make sure that you're paying attention to the right things, so that you can avoid unnecessary confusion and so that your hard work will pay off.

STRUCTURE OF THE BOOK

This book is split into two parts. Part 1 is about mathematical content and Part 2 is about the process of learning.

Part 1 could be called "Things that your mathematics professor might not think to tell you." It describes the structure of advanced mathematics, discusses how it differs from earlier mathematics, and offers advice about things you could do to understand it. I've put this part first because it is probably what is expected by students who read this before declaring their major. Note that I do not aim to teach the mathematics—that is the job of your professors and instructors. So you will not find that the book contains a lot of mathematical content. What it contains instead is information on how to *interact* with the content. It thus includes detailed illustrations, but not exercises. If you want exercises, there are many good books you can work with, and I list some of these in the Further Reading section at the end of each chapter.

Part 2 is about how to get the most out of your lectures, and about how to organize yourself so that you can keep up with the mathematics and therefore enjoy it. I've put this part second because I expect that many students who are relatively new to college studies will be thinking, "Pah! I don't need information on study skills! I have done well in dozens of exams already and clearly I am a good student." If that is what you're thinking, good for you. But maybe Part 1 will convince you that, because the nature of the mathematics changes as you move into upper-level courses, some tweaks to your approach might be useful. Indeed, some people might be reading this book precisely because they know that they are not organizing their studies very well. Those in this position might want to read Part 2 first. And anyone who is so far behind that they find themselves in a state of panic should turn straight to Chapter 12 and start there.

To fit in with the experience of all readers, I decided to write as though I'm addressing someone who is on the point of declaring their major, and to present material that should be interesting but challenging to someone in that position. This means that you might encounter new concepts while you are reading, and you'll certainly have to think hard. I have done my best to explain everything clearly and, as I mentioned above, you should be willing to be challenged as an undergraduate student. But you might find it useful to come back to some of the ideas later, when you have more experience to draw on. I hope that this book will be useful throughout your mathematics major.

USEFUL VOCABULARY

I have tried to make each chapter fairly self-contained, so that you can jump in anywhere (though that was harder with Part 1, so I do recommend that you read most of that in order). I have also tried to introduce technical terms when they are needed, and have provided a list, prior to this introduction, summarizing the mathematical notation used in the book. However, you will find it useful to be familiar with the following vocabulary for some practical aspects of undergraduate studies.

Mathematics major: In their first two years, intending *mathematics majors* usually take a sequence of Calculus courses and some Linear Algebra. They are usually required to declare a major in their second year. Some colleges require new majors to take a "transition" or "bridging" course to prepare them for upper-level mathematics by introducing ideas about proof. Upper-level work usually includes Real Analysis and Abstract Algebra, but most institutions offer a variety of courses in both pure and applied mathematics, and students usually have some choice about which ones to take. This book is designed to be useful for all mathematics majors, although it focuses on the transition to upper-level courses and on pure mathematics.

Sections: Students electing to take a course will have to register for a particular *section*. For lower-level courses that are taken by hundreds of students every semester, there might be many sections to choose from. For upper-level courses on more specialist topics, there might be just one or two. Different sections will cover the same material but will be taught at different times and places and (often) by different professors. In institutions that operate honors programs, there might be specific honors sections; students wishing to register for these might need some form of special permission.

Recitations and problems sessions: Lots of courses have associated *recitations* or *problems sessions*; sometimes a large lecture class might be split into multiple smaller classes for recitations. These sessions focus on working through problems, and they might be run by the course professor or by an alternative instructor such as a postgraduate teaching assistant.

Problem sets: Mathematics professors distribute sets of problems for their courses. These might be lists of exercises from a course textbook, or they might be on separate printed sheets produced by the professor. Problem sets are usually distrubuted weekly, or perhaps when a topic or chapter begins or ends. I'll refer to them as *problem sets*, but they are also variously called *example sheets*, *exercises*, or just *homework*.

Coursework assignments: Mathematics professors also set work that is to be submitted for credit; that is, to contribute to the grades that students are eventually awarded. This work takes many forms, including selected problems from problem sets, separate written assignments, and tests taken during class time or on a computer. I'll refer to these things collectively as *coursework assignments*.

Online materials: Colleges usually have a computer environment via which students can access academic materials. In some cases this will be a *virtual learning environment* (VLE), which might be referred to by a commercial or locally-decided name (at my university, the VLE is called "Learn"). You usually have to log in to a VLE, but when you do you'll find that the system knows what courses you are registered for and provides a link to a separate page for each one. Within each course page, your professor can provide resources (lecture notes, problem solutions, links to online tests, and so on) for you to download. Alternatively, there might be no VLE as such, but course webpages maintained by your department or by individual professors. The main college pages will also have links to other services and facilities, and might allow you to make college-related financial transactions. Needless to say, someone will provide you with information about all of this.

Advisers: US-style college systems give students a lot of choice about how to organize their personal degree programs; about how to select courses to make good progress through a major (and minor) while also fulfilling other requirements. Prerequisite structures mean that this is not a trivial process, so *academic advisers* are usually available to help students make sensible decisions.

One final term worth mentioning is *independent learning*. Students sometimes misinterpret this term. They know that independent learning is expected in college, but some think this means that they will have to study alone with no help or support. This is far from true, as I hope will be clear from the whole of this book. There will be challenges, and you will have to put in some individual intellectual effort. But those with a bit of initiative will find that plenty of help is available (see especially Chapter 10) and that asking good questions and thinking in productive ways can lead to rapid progress.

With that in mind, let's get started.

PART I
Mathematics

CHAPTER 1

Calculation Procedures

This chapter addresses some issues that frequently arise when students make the transition to advanced mathematics. It discusses ways in which undergraduates can build on their existing mathematical skills; it also identifies ways in which mathematical expectations change when a student moves from lower-level to upper-level courses. It describes different approaches to learning, and argues that certain approaches are more useful than others when dealing with advanced mathematics.

1.1 Calculation and advanced mathematics

The first part of this book is about the nature of upper-level mathematics. Upper-level mathematics has much in common with lower-level mathematics, and students who have been accepted onto a mathematics major already have an array of mathematical skills that will serve them well. On the other hand, upper-level mathematics also differs from lower-level mathematics in some important respects. This means that most students need to extend and adapt their existing skills in order to continue doing well. Making such extensions and adaptations can be difficult for those who have never really reflected on the nature of their skills, so Part 1 discusses these in some detail.

One thing you have certainly learned to do is to apply mathematical procedures to calculate answers to standard questions. Some people enjoy doing this type of work. They like the satisfaction of arriving at a page of correct answers, and they like the security of knowing that if they do everything right then their answers will, indeed, be correct. Sometimes they compare mathematics favorably with other subjects in which things seem to be more a matter of opinion and "there are no right answers."

Other people dislike this aspect of mathematics. They find it dull to do lots of repetitious exercises, and they get more satisfaction from learning about *why* the various procedures work and how they fit together. I will discuss this difference further in this chapter. For now, however, note that knowing how to apply procedures is extremely important because, without fluency in calculations, it is hard to focus your attention on higher-level concepts.

When you start taking upper-level courses, professors will expect you to be fluent in using the procedures you have already learned. They will expect you to be able to accurately manipulate algebraic expressions, to solve equations, to differentiate and integrate functions, and so on. They will expect you to be able to do these things without having to stop each time to look up a rule, and they might not be patient with students who are not able to do so. This is not because they are impatient with students in general—most professors will be very happy to spend a long time talking with you about new mathematics, or responding to students who say, "I know how to do this, but I've never really understood why we do it this way." But they will not expect to have to re-teach things you have already studied. So you should brush up your knowledge prior to beginning a course, especially if, say, you've done no mathematics all summer.

Once you do begin, you'll find that some upper-level mathematics involves learning new procedures. These procedures, unsurprisingly, will be longer and more complicated than those you met in earlier work. I am not worried about your ability to apply long and complicated procedures, however, because to have got this far you must be able to do that kind of thing. Here, I want to focus on more substantive changes in the ways in which you have to interact with the procedures.

1.2 Decisions about and within procedures

The first substantive difference is that you will have more responsibility for deciding which procedure to apply. Of course, you have learned to do this to some extent already. For instance, you have learned how to multiply out brackets and write things like:

$$(x + 2)(x - 5) = x^2 - 3x - 10.$$

But hopefully you have also learned that it is *not* sensible to multiply out when trying to simplify a fraction like this:

$$\frac{x^2(x + 2)(x - 5)}{x^2 + 2x}.$$

For the fraction, simplification is easier if we keep the factors "visible." Nonetheless, many people automatically multiply out, probably because multiplying out was one of the first things they learned to do when studying algebra. They would, however, become more effective mathematicians if they learned to stop and think first about what would allow them to make the most progress. If this doesn't apply to you for this particular type of problem, does it apply in others? Have you ever done a long calculation and then realized that you didn't need to? Could you have avoided it if you'd stopped to think first? Part of deciding which procedure to apply is giving yourself a moment to think about it before you leap in and do the first thing that comes to mind.

This might not sound like a big deal, but think for a moment about how often you *don't* have to make a choice about what procedure to apply. Often, questions in books or on tests tell you exactly what to do. They say things like, "Use the product rule to differentiate this function." Even when a question doesn't tell you outright, it is sometimes obvious from the context. In high school, if your teacher spent a lesson showing you how to apply double angle formulas, then gave you a set of questions to do, it was probably safe for you to assume that these would involve double angle formulas. This helped you out, but it means that much of the time you didn't have to decide what procedure to apply. In the wider world, and in advanced mathematics, making decisions is more highly valued and more often expected. This means that questions presented to you on problem sheets or in exams will usually just say "solve this problem" rather than "solve this problem using this procedure."

Another part of deciding which procedure to apply is being able to distinguish between cases that look similar but are best approached in different ways. For example, consider integration, and more specifically integration by parts. You may know that this is used when we want to integrate a product of two functions, one of which gets simpler when we differentiate it, and the other of which does not get any more complicated when we integrate it. For instance, in $\int xe^x \, dx$, x gets simpler if we differentiate it, and e^x gets no more complicated if we integrate it. You might also know, however, that sometimes mathematical situations look superficially similar, but are best tackled using different procedures. In the integration case, integration by substitution might sometimes be more appropriate. For instance, in $\int xe^{x^2} \, dx$, we would probably want to use integration by substitution instead of by parts. Can you see why?

Integration by substitution is a good case for another point I want to make, this time about making decisions *within* procedures. It might be that you read the end of the last paragraph and thought, "But what substitution should I use?" Perhaps your teachers or books always told you what to use, but I would argue that they shouldn't necessarily have to. After a while, you should notice that certain substitutions are useful in certain cases. If you pay attention to the structures of these cases then, even if you wouldn't be sure that you could pick a good substitution for a new case, you should have an idea of some sensible things to try. If you haven't deliberately thought about this before, I suggest you do so now. Get out some questions on integration by substitution and, without actually doing the problems, look at the suggested substitutions. Can you anticipate why those substitutions will work? Can you then anticipate what would work in similar cases? I'll come back to this illustration later in this chapter.

So how can a student improve their ability to make decisions about and within procedures? I have two suggestions. The first is to try doing exercises from a source where the procedure to be applied is not obvious. A good place to look is in end-of-chapter exercises, which tend to cover more material. The second suggestion is to turn ordinary exercises into opportunities for reflection. When you finish an exercise, instead of just moving on to the next one, stop and think about these questions:

1. Why did that procedure work?
2. What could be changed in the question so that it would still work?
3. What could be changed in the question so that it would *not* work?
4. Could I modify the procedure so that it would work for some of these cases?

All of these questions should help you build up flexibility in applying what you know.

1.3 Learning from few (or no) examples

When you learned a new procedure in high school, your teacher probably introduced it by showing you several worked examples. These examples probably varied a bit, so that the first ones were easier and the later ones were harder. Your teacher probably then set you some exercises so you could practice for yourself. You probably did these exercises with the worked examples to hand, applying the method to the new cases you were given.

In upper-level courses it's less common to have several worked examples to hand when you begin trying a problem. You might have just one or two. These will not encompass all the possible variation in applying the procedure, so you will be more responsible for working out whether you can follow it exactly, or whether you need to adjust it because you're working with a slightly different situation.

For a straightforward example in which an adjustment is necessary, consider school students who have learned to solve quadratic equations like $x^2 - 5x + 6 = 0$ by factorizing then setting the factors equal to 0. They might write something like this:

$$x^2 - 5x + 6 = 0$$
$$(x - 2)(x - 3) = 0$$
$$x - 2 = 0 \text{ or } x - 3 = 0$$
$$x = 2 \text{ or } x = 3.$$

Now suppose such a student is asked to solve the equation $x^2 - 5x + 6 = 8$, and writes this:

$$x^2 - 5x + 6 = 8$$
$$(x - 2)(x - 3) = 8$$
$$x - 2 = 8 \text{ or } x - 3 = 8$$
$$x = 10 \text{ or } x = 11.$$

What exactly has gone wrong here? Make sure you can answer this—spotting the errors in logical arguments is important. Can you explain what the error is and why it is an error? Can you see, nonetheless, why someone might make this mistake? Notice that it isn't crazy—the procedure looks on the surface like it might work because the equation seems to be of the same kind. It *doesn't* work in this case because, while it is true that if $ab = 0$ then we must have $a = 0$ or $b = 0$, it is *not* true that if $ab = 8$ then we must have $a = 8$ or $b = 8$. A sensible modification of the procedure would have been to subtract 8 from both sides first, and work with an equation in the standard form.

This is a simple illustration and, when you learned to do this kind of thing, your teachers probably didn't expect you to use just one worked example to work out how to deal with related but non-identical cases. Indeed, in lower-level mathematics, students are usually only shown cases in which everything works—comparatively little time is devoted to recognizing examples for which standard procedures do not apply. So you might not have had much practice at the critical thinking needed to spot the limitations of procedures, and you should be prepared to develop this skill as you work toward your major.

In fact, in upper-level mathematics, you will sometimes be asked to apply a procedure without having seen *any worked examples at all*. This might sound impossible, but it isn't, because useful information sometimes comes in other forms. In particular, applying procedures often involves substituting things into formulas. For instance, the product rule for differentiation involves a formula, and is often expressed as follows (if the formula you use does not look exactly like this, can you see how it corresponds to the one you're familiar with?):

$$\text{If } f = uv \text{ then } \frac{\mathrm{d}f}{\mathrm{d}x} = u\frac{\mathrm{d}v}{\mathrm{d}x} + v\frac{\mathrm{d}u}{\mathrm{d}x}.$$

To apply the product rule, we decide what we want u and v to be, then work out all the other things we need and substitute them into the formula. When you first saw this, your teacher probably went through a few examples, showing you how to do this. Was that really necessary, though? If you learned something similar now, would you need someone to walk you through it step by step, or could you just make all the appropriate substitutions and follow it through yourself?

For another illustration, consider the definition of the derivative. You may have seen this when you were first introduced to differentiation. You'll have to engage with it in detail during your major, so we'll use it here to show how you might apply a general formula without needing a worked example. Here it is:

Definition: $\dfrac{df}{dx} = \lim_{h \to 0} \dfrac{f(x+h) - f(x)}{h}$, provided this limit exists.

The question of why this is a reasonable definition is important, as is the question of why we need to say "provided this limit exists." However, we're not focusing on those questions here (look out for the answers in a course called Advanced Calculus or Analysis). For now, suppose we've been given this definition and asked to use it to find the derivative of the function f given by $f(x) = x^3$ (I know you already know what the answer is, but humor me for a moment). What would we do? Well, we want to find df/dx, so the formula can be used exactly as it is—we don't need to do any rearranging. We have a formula for f, so we can substitute that in:

$$\frac{df}{dx} = \lim_{h \to 0} \frac{(x+h)^3 - x^3}{h}.$$

Then we can simplify the resulting expression:

$$\begin{aligned}
\frac{df}{dx} &= \lim_{h \to 0} \frac{(x+h)^3 - x^3}{h} \\
&= \lim_{h \to 0} \frac{x^3 + 3x^2h + 3xh^2 + h^3 - x^3}{h} \\
&= \lim_{h \to 0} \frac{3x^2h + 3xh^2 + h^3}{h} \\
&= \lim_{h \to 0} 3x^2 + 3xh + h^2.
\end{aligned}$$

Finally, we can observe that as h tends to 0, $3x^2$ stays as it is, but $3xh$ tends to 0 and so does h^2. So the limit is equal to $3x^2$. Hence, by substituting into the definition, we have established that

$$\frac{df}{dx} = 3x^2.$$

This is what we were expecting.

Recognizing that you can apply general formulas directly to examples, on your own, should make you feel mathematically empowered. It might still be reassuring to have someone walk you through some worked examples, but in many cases you don't need that level of support any more.

1.4 Generating your own exercises

Sometimes, as I said, university professors will give you only a small number of worked examples. Similarly, sometimes they will set you only a small number of exercises. They might, for instance, demonstrate how we can prove that the function f given by $f(x) = 2x$ is continuous,[1] and ask you to write a similar proof for $f(x) = 3x$. They might not ask for anything more, but their intention will be that you can then see how to write a similar proof for $f(x) = 4x$ and for $f(x) = 265x$, and so on. Even if you can, you might be well-advised to write out a proof for a few cases anyway, just for practice. In high school, your teacher probably took responsibility for deciding how much practice you should do, but a professor is more likely to set just one exercise and leave it to you to judge whether you'd benefit from inventing similar ones.

You might also be well-advised to stop and ask yourself about the limitations of such a proof. Would it, for instance, work for negative values? Would the same steps apply immediately for $f(x) = -10x$, or would you need to make some kind of adjustment in that case? What about for $f(x) = 0x$? Indeed, could you generalize properly, and write an argument for $f(x) = cx$? Would you need to place any restrictions on c? Mathematicians consider this kind of thinking to be very natural. They probably always did it for themselves, without having to be told to. You should too.

The upshot of this chapter so far is that you will not succeed in upper-level mathematics if you always try to solve problems by finding something that looks similar and copying it. In some cases, if you copy without

---

[1] If you can't see why there would be anything to prove, wait for the discussion in Chapter 5.

thinking, you could end up writing nonsense because some property that holds in the worked example does not hold in your problem. In other cases, copying might just be inefficient—the first approach you choose might work perfectly well but take twice as long as a different method. In still other cases, there might not be any examples to follow at all. You might have to recognize that a particular definition or theorem can be applied, then apply it, going straight from the general statement to your specific case via appropriate substitutions.

This makes upper-level mathematics more demanding than mathematics you have encountered before. You have to attend more carefully to whether your symbolic manipulations are valid for the example that you are working with. This is not always easy, but again you can practice by asking yourself the questions at the end of Section 1.2.

1.5 Writing out calculations

If you've read the Contents pages of this book, you will have noticed that there is an entire chapter devoted to writing mathematics. I'm not going to say much about this here, but I have a few comments that are particularly relevant while we're talking about calculation procedures.

There are probably some procedures for which you can happily keep everything in mind and write down only minimal working, but others for which you tend to make errors so that it is worth doing one step at a time and writing everything down. If you have been told that you should *always* write out all of your working, this would be a good time to begin letting go of that idea. One of the great things about mathematics is that it is very compressible: we can understand complex ideas by mentally "chunking" their components. For example, when expanding brackets, you probably originally learned to write everything out, like this:

$$(x + 2)(x - 5) = x^2 + 2x - 5x + 2(-5) = x^2 - 3x - 10.$$

But now you probably do much of it in your head and just write this:

$$(x + 2)(x - 5) = x^2 - 3x - 10.$$

Doing so helps you to do lots of calculations faster, which helps you to keep your eye on whatever overall problem you are trying to solve.

At lower-levels, it may be in your interest to write out a certain amount of working in order to satisfy an examiner, but you can still think for yourself about what you actually need. At upper-levels, you will find that your professors sometimes take routine calculations for granted; they will "skip steps" and expect you to be able to fill them in for yourself. For instance, to go back to integration by substitution, I am happy to just write down:

$$\int x \cos\left(x^2\right) \mathrm{d}x = \frac{1}{2} \sin\left(x^2\right) + c.$$

I can do the integration quickly in my head because I know that I am looking for something that differentiates to $x \cos\left(x^2\right)$. I know that the answer will be something like $\sin\left(x^2\right)$ so I'd write that down first. I can then see that if I differentiate $\sin\left(x^2\right)$ (using the chain rule), I get $2x \cos\left(x^2\right)$, so I just need to divide the $\sin\left(x^2\right)$ by 2 to get what I want. If you haven't done so already, you might try using similar thinking for the integral $\int x e^{x^2}\, \mathrm{d}x$ as mentioned earlier in this chapter.

Either calculation can be written out fully using the substitution $u = x^2$ (if you're unsure about this, try it, and you'll find that what you write is more or less exactly the reasoning I just went through). If I were lecturing a freshman Calculus course, I might expect a student to write everything out. If I were lecturing anything other than that, however, I would expect my students to be able to do this kind of thing in their heads. I probably wouldn't mind if they just wrote down the answer; chances are they'd be doing it as part of some larger problem, and their solutions to the larger problem would be more concise if they chose not to include every last detail. It's certainly likely that if I needed this calculation in a lecture, I would just write down the answer and expect students to be able to check its correctness for themselves. If you get used to seeing mathematics as compressible, you should be fine with this.

I should stress, however, that your teachers haven't said anything wrong if they've told you to write out your working in full. Doing so certainly has benefits: it allows you to check your work for errors and to remember what you were thinking when you come back to a problem. But compression is also important, as is making judgments about what will make your overall argument clearest to a reader. We will return to these themes throughout the book.

1.6 Checking for errors

If you've got this far in mathematics, you have probably made lots of mathematical errors. Some of these will have been small, like missing a constant or accidentally writing a "+" instead of a "−". Some you will have noticed straightaway, others you will have been confused by for much longer. When I was in high school, for instance, I was once totally mystified about why a simple piece of algebra wasn't coming out right. My teacher let me look at it for a couple of days before he laughed (not unkindly) and pointed out that I was adding where I should have been multiplying. The feeling that I had been a bit daft was greatly outweighed by relief that I had not gone mad and failed to understand the whole problem in some profound way. At any rate, you are probably in the habit of checking your work for minor manipulation errors, which is good.

Some errors, however, are more serious. In our example about factorizing the quadratic, the error indicates that the person has failed to understand why the method works. You should be on the lookout for such problems, and check that each step really works in the way that you think. Of course, you might try to do that, but still not get the expected answer. When that happens, don't be afraid to talk to someone about it. A lot of the queries I get from students are along the lines of "I know this isn't working but I can't find the mistake." They often learn a lot when we sort it out, usually because they recognize that they were making an assumption without being aware of it. Increased awareness of some property or principle means that you can recognize it in more situations and can use it actively in further thinking.

Other errors are not serious but are annoying to professors because they look sloppy. These occur when students give solutions that they should have recognized could not possibly be right. If the answer is a number, perhaps it is way too large or way too small. One common source of errors like that is the calculator. Calculators, like all computing devices, are fast but stupid. They will give you the answer to the question you asked, whether or not this is the question you intended to ask. If you hit the wrong key, or you don't quite know how your calculator handles certain kinds of input, or you are working in degrees when you want radians, your calculator won't know that. It will dutifully give you what it thinks you asked for. Only you can judge whether the output is plausible.

But this sort of thing can happen in other ways too. Sometimes students give an answer that is *not even the right type of mathematical object*. They give a number when the answer should be a vector, or a function when the answer should be a number, or something like that. This can happen in pretty straightforward ways: say that students are asked to find $f'(2)$ and they find $f'(x)$ (a function) but forget to substitute in 2 for x to find the final answer (a number). It can also happen in complex ways, and is sometimes an indication that a student hasn't really understood a question, and has copied a procedure that looks superficially similar but does not achieve what is required. I'll talk about recognizing and avoiding that kind of mistake in Chapter 2.

In the meantime, here's a piece of advice I give to students, especially for exams. If you have done some calculations and you know your answer must be wrong, write a quick note to indicate this: something like "must be wrong because too small." At least then the person grading your work knows that you were thinking about the meaning of the question. Obviously, it's better if you can indicate where the mistake has happened, and even better if you can fix it. But, if you can't do those things in the time available, you can at least demonstrate that you know what a reasonable answer would look like.

1.7 Mathematics is not just procedures

This chapter has been all about calculational procedures. But I want to come back to the idea that knowing how to apply such procedures is only one part of understanding mathematics. It is an important part, but most people can recognize the difference between learning to apply a procedure mechanically, and understanding why it works. Learning mechanically has some advantages: it is generally quick and relatively straightforward. But it also has disadvantages: if you learn procedures mechanically, it is easier to forget them, to misapply them, and to mix them up. Developing a proper understanding of why things work is generally harder and more time-consuming, but the resulting knowledge is easier to remember and more supportive of flexible and accurate reasoning.

Whatever your previous experience, you almost certainly have some things that you understand deeply and some things that you have only learned procedurally. For instance, you could probably explain why we "change signs" when we "move the 5 to the left" in order to solve the equation $x + 3 = 2x - 5$. You might explain that what we're really doing is adding 5 to both sides and that, because the two sides were equal before and we've done the same thing to both, they're still equal now. This demonstrates good understanding because you know not only *that* we do certain things, but also *why* these things are reasonable. You can give an explanation that's genuinely mathematical, and is thus better than "that's what it says in the book" or "that's what my teacher told me."

It's hard to claim a "full" understanding of something, because there are so many links between various pieces of mathematics that it would be tough for anyone to say that they are familiar with all of them (especially in a world where new mathematics is being developed all the time—see Chapter 14). For example, if you followed the discussion about the definition of the derivative, you should be able to explain why the derivative of $f(x) = x^3$ is $f'(x) = 3x^2$. This would be a good explanation but it wouldn't, on its own, tell us how this result relates to the graphs of the functions, or why the derivative of $f(x) = x^n$ should in many other cases be $f'(x) = nx^{n-1}$. It also doesn't tell us why mathematicians use that definition in the first place. (Perhaps you know why. If not, think about it, look it up somewhere, and look out for it in your Calculus courses.)

In fact, you should beware of being too confident about your understanding, even for straightforward mathematics. There might be things that you've known for a long time and can use confidently, but that you can't explain as well as you'd think. For instance, you know that when we multiply two negative numbers together we get a positive one. But do you really know why? Could you give an explanation that would convince a skeptical 13 year-old? Similarly, you know that $5^0 = 1$, but why is this the case? You know that we "can't divide by 0" but, again, why? Why can't we just say that $1/0 = \infty$? If you are tempted to answer any of these questions by saying "that's just how it is," be aware that there are reasons for all of these things, even if you don't yet know them. You might like to see the further reading suggestions to find out how professional mathematicians would answer.

In the meantime, there will be other things that you know you only understand procedurally. Perhaps you can use the quadratic formula, but you'd be hard pressed to explain why it works (we'll look at this in Chapter 5). Perhaps you're good at integration by parts, but you haven't the faintest idea where the formula comes from. Perhaps you can work with the formulas for simple harmonic motion, but you don't really know what the symbols are telling you or why it's reasonable to use those equations to capture that type of motion. One great thing about advanced mathematics is that you can expect explanations of many such formulas and relationships.

Indeed, plenty of mathematicians will tell you that learning procedures mechanically is bad—that you should always be striving for a deep understanding. This is a well-intentioned claim, but it is a bit unrealistic. For a start, there are plenty of situations in which good understanding is not really accessible to you at a given stage. Limits, for example, appear in some Calculus courses, but are treated fairly informally. In upper-level courses, you will learn a formal definition of limit, and you will learn how to apply it and how to use it to prove various theorems. But the definition is logically complex and, if you are like most people, it will cost you some effort to learn how to work with it. When you can, you will recognize that your understanding has improved, but you will probably think that your professors were right not to introduce it earlier.

The claim is also unrealistic because mechanical knowledge of procedures can be very useful. The classic real-life examples are driving a car or operating a computer. Probably you can do at least one of these things, but you have no real idea how an internal combustion engine works or how a computer turns your keystrokes into letters that appear on the screen. You could acquire that understanding, but you'll likely do just fine in life if you don't. In mathematics, there are many analogous situations. Sometimes we don't learn about the details of something for pragmatic reasons—there just isn't time. Sometimes we don't learn about them because they are fiddly and because studying them would distract us from understanding what a procedure achieves. There will still be situations like this during your major, but you will start to see more and more of the theory underlying mathematical knowledge.

Sometimes people don't try to develop deep understanding because they don't need it for the tasks they want to accomplish. Engineering students famously get annoyed when mathematics professors try to promote understanding of why procedures work; they say "Just tell us what to do!" If you're of a practical bent, you will probably be tempted to say something similar, at least in your pure mathematics courses. In applied mathematics (mechanics, statistics, decision mathematics, and so on), there does tend to be more focus on solving problems and less on developing abstract theories. But, even in those subjects, you will find that mathematicians care very much that you should understand why your calculations are reasonable. This is partly so that you can avoid making mistakes like those discussed in this chapter, partly so that you can be more flexible in adapting procedures to new problems, and partly just for the joy of it.

I certainly think that, wherever possible, you should aim for a deep understanding of why mathematical procedures work and why mathematical concepts are related to each other as they are. Such understanding is more powerful and more memorable, and acquiring it is hard work but very satisfying. Nonetheless, I expect that you will end up learning some things procedurally, either because that's the only way that's accessible, or because you're not really interested in a particular subject but you do want to pass the exam. My view is that, as long as you are aware that you might be vulnerable to certain kinds of error, that's fine. As with many suggestions in this book, this amounts to taking responsibility for your own learning.

SUMMARY

- Before taking upper-level courses, you should brush up your knowledge of standard procedures because professors will expect you to be able to use these fluently.
- As you progress, you will be expected to take more responsibility for deciding which procedure to apply; it might be a good idea to practice this by working on exercises from sources that do not tell you exactly what to do.
- You will also be expected to adapt procedures in sensible ways, and to work out how theorems or definitions can be applied without necessarily having seen many worked examples.
- You might be given only a small number of exercises; it might be sensible to invent your own similar ones for practice.
- You will not succeed in upper-level mathematics if you always try to solve problems by finding something that looks similar and copying it. You have to be more thoughtful than that.

- It is not always necessary to write out calculations in full. Your professors will sometimes skip steps, and you might want to do the same if it makes an overall solution clearer.
- Try to avoid calculator errors by checking whether your answer is plausible given the problem situation. Also, think about what type of object (a number, a function, etc.) you are expecting your answer to be.
- Mathematics is not just about procedures. Fluency with procedures is important, but in many cases you should aim for a deeper understanding of why procedures work.

FURTHER READING

For a guide to becoming a more effective mathematical problem solver, try:

- Krantz, S. G. (1997). *Techniques of Problem Solving*. Providence, RI: American Mathematical Society.
- Mason, J., Burton, L., & Stacey, K. (2010). *Thinking Mathematically (2nd Edition)*. Harlow: Pearson Education.
- Pólya, G. (1957). *How to Solve It: A New Aspect of Mathematical Method (2nd Edition)*. Princeton, NJ: Princeton University Press.

For insights into how mathematicians think in a sophisticated way about numbers and arithmetic, try:

- Gowers, T. (2000). *Mathematics: A Very Short Introduction*. Oxford: Oxford University Press.

For ways of thinking more deeply about school mathematics, try:

- Usiskin, Z., Peressini, A., Marchisotto, E.A., & Stanley, D. (2003). *Mathematics for High School Teachers: An Advanced Perspective*. Upper Saddle River, NJ: Prentice Hall.

For more on childrens' understandings of mathematical concepts, try:

- Ryan, J. & Williams, J. (2007). *Children's Mathematics 4–15: Learning from Errors and Misconceptions*. Maidenhead: Open University Press.

CHAPTER 2

Abstract Objects

This chapter explains how to think of mathematics in terms of abstract objects. Some abstract objects, such as numbers and functions, will be familiar; others, such as binary operations and symmetries, will not. This chapter explains why it is important to be able to think of concepts as objects that are organized into hierarchical structures. It points out things that can go wrong for students who do not master this, and highlights some ways of thinking that are particularly useful when studying abstract pure mathematics courses.

2.1 Numbers as abstract objects

This chapter is about abstract mathematical objects. Take the number 5, for instance. When you first started dealing with the number 5, it was probably in the context of counting sets of things (oranges, wooden blocks, unifix cubes, or whatever). 5 was thus associated with the process of saying "1, 2, 3, 4, 5," and pointing at these things. At some point, however, you stopped needing to point at actual physical objects. In fact, you stopped needing to think about five *of* something at all, and became able to think about 5 as a thing in its own right. This sounds innocuous but it is an incredible feat of the human intellect, and one that is vital to success in mathematics because it means that you can divorce the abstract object 5 from the potentially lengthy process of counting.

To see why that's so significant, think about all the things you know about 5. You know that $5 + 2 = 7$, that $5 + 5 = 10$, that $5 \times 5 = 25$, and so on. Did you think about any physical objects when you were reading those? Even if you did, it probably wasn't 25 oranges. And how about this: you can also use your 5-related knowledge to quickly derive other results that you have never thought about before. You can assert with total confidence that $6015 + 5 = 6020$, for instance. You've probably never thought about that particular sum before, but you can do so very easily. And I bet you didn't think about 6020 oranges. The number 5 gets its power not only from referring to sets of five physical things, but also from the fact that it interacts with other numbers in highly predictable ways.

The reason this is so important is that counting takes time. It is much slower than just manipulating facts about number relationships. Finding $6015 + 5$ by counting would take ages. You probably wouldn't even want to count for $13 + 18$, say, because it would take a while and because you know you would be error-prone. If you had to count, because you didn't know many arithmetic facts or because you couldn't easily derive new facts from them, you would probably come to think that arithmetic was impossible. This sounds far-fetched, but research in mathematics education reveals that this is exactly what happens. When young pupils do not do well in arithmetic, it is often because they fail to notice and build on these regularities. Instead, they try to do most of their arithmetic by counting. They try to find $13 + 18$ by counting to 13 and counting to 18 and then counting the whole lot. And this isn't the worst case: imagine trying it for $31 - 14$. A student who sees $31 - 14$ as an instruction to count has a lot of work on their hands. Counting quickly becomes impossibly cumbersome, and to make progress it is critical to stop thinking of 5 as an instruction to count, and start thinking of it as an object that interacts in certain ways with other numbers and operations.

Those who succeed do notice these arithmetic regularities; they build up a large and highly interconnected system of known facts from which they can rapidly and confidently derive new ones. Those who study undergraduate mathematics also cope with the introduction of fractions, decimals, functions, algebra, and a whole host of other things that confuse many students when they first appear. People in your position are thus able to think about objects that are much more abstract than numbers, and this skill will serve you well in upper-level courses. Unsurprisingly, however, you will have to do it more rapidly, more flexibly, and with considerably more complex kinds of objects. This chapter is about how to do so effectively.

2.2 Functions as abstract objects

We'll start with another concept that can be thought of in different ways: functions. When you first met functions, it was probably as "machines" that take inputs and do something to them to produce outputs. Your job was to take a given input (say, 6), "perform" the function (say, multiply by 2 and add 1) and report the answer. You needed to be able to take a simple input and follow some instructions, but that's it.

Later, you were asked to do more sophisticated things. At some point you were probably asked to look at a table of inputs and outputs and work out what the function could be. To do this you had to know that you were supposed to assume the existence of some underlying process that does the same thing to every input, and you had to think about what that process could be. You might have been asked to find inverse functions, too, which again demands being able to conceive of a function as a process, this time in order to imagine reversing it.

Finally, you learned to do things that demand thinking of a function, not as a process sending numbers to other numbers, but as an object in its own right. In some senses this is fairly natural. Linguistically, for instance, we talk about "**the function** sine x." In such phrases, the function is a noun—the language treats it as an object. Graphically, it is quite natural too. We're accustomed to seeing a graph of the sine function, and this representation makes it natural to think of the function as a single unified object.[1]

In other senses, treating a function as an object might feel less natural. Consider, for instance, the process of differentiation. Differentiation takes a function as an input, does something to it, and gives another function as an output. For instance, if we take the function given by $f(x) = x^3$ and differentiate it, we get the function $f'(x) = 3x^2$. If we take the function given by $g(x) = e^{5x}$ and differentiate it, we get the function $g'(x) = 5e^{5x}$. In this sense, differentiation can be thought of as a higher-level process.

[1] By the way, when you first met the idea of the sine of an angle, you probably spent some time measuring lengths and calculating sines of particular angles. When we talk about sine as a function, we're talking about the function that takes each angle to its sine, so we're dealing with all of those possible sine calculations at once. This is similar to, say, polynomial functions, except that as well as treating the input as a number, we can also think of it as an angle.

Where a function would take numbers as inputs and return numbers as outputs, differentiation takes functions as inputs and returns functions as outputs. You might find it helpful to look out for other cases where processes are treated as objects, and we'll come back to this idea later in the chapter.

2.3 What kind of object is that, really?

Whatever mathematics you've studied so far, you will be familiar with a variety of different types of mathematical object. As well as numbers and functions, you might know about vectors, matrices, complex numbers, and so on. You're probably very flexible in thinking about these. You probably know about all sorts of relationships and associated procedures. But how flexible are you in thinking about these objects *as objects*? We'll think about this before we move on.

First ask yourself, what kind of object is 10? This is not a trick question. It's a number, obviously. Now, is it a fraction? If your instinct is to say "no," stop and think about how you would answer this alternative question: Is a square a rectangle? To this second question, lots of young children say "no"—you can probably see why. However, a square is a rectangle because it has all the properties it needs in order to be a rectangle. It happens to have extra properties too but, mathematically, that's irrelevant to the question. Now, can you see the analogy between this and the question about 10? We don't normally write 10 as a fraction, but we easily could. It's 10/1, for a start. Or it's 20/2, or 520/52, or any number of other things. It's a whole number as well, but again this means that it has the properties needed to be a fraction, plus some extra ones.

To state this as you'll hear it in upper-level courses, 10 is an *integer*. *Integer* is the proper mathematical word for "whole number," and the set of all integers is denoted by $\mathbb{Z}$, which comes from the German word "Zahlen," meaning numbers. We write $10 \in \mathbb{Z}$ to mean "10 is an element of the set of integers." (10 is a *natural number* too,[2] where the natural numbers $1, 2, 3 \ldots$ are denoted by $\mathbb{N}$.) 10 is also *rational number*, where

---

[2] Some people include 0 in the natural numbers and some do not, so make sure you know what your professors are including.

rational numbers are those that can be written in the form p/q, where both p and q are integers (and q is not zero).[3] We just saw some ways to write 10 in this form. The set of all rational numbers is denoted by $\mathbb{Q}$, which comes from the word "quotient." In mathematical terms, we say that $\mathbb{Z}$ is a *subset* of $\mathbb{Q}$, which means that everything in $\mathbb{Z}$ is also in $\mathbb{Q}$ ($\mathbb{Q}$ contains a lot more stuff too, of course). We sometimes write this as $\mathbb{Z} \subseteq \mathbb{Q}$. Notice that the $\subseteq$ symbol is like the $\leq$ symbol but curvy. Can you see why this is a sensible choice of notation? Mathematicians like this kind of regularity.

So, 10 is a rational number. Now, is it a complex number? Yes. 10 can be written as $10 + 0i$. It's a perfectly good complex number—it just happens to be an integer, and a rational number, and a real number as well. By the way, the notation for the set of all complex numbers is $\mathbb{C}$, and the notation for the set of all real numbers is $\mathbb{R}$. No big surprises there.

Now, this kind of thing happens quite a lot. Often we can think of objects as being of one type or another depending on which is more illuminating at the time. Furthermore, analyzing object types can clarify the meaning of different collections of symbols. For instance, what types of object are these?

$$\int_1^5 3x^2 + 4 \, dx, \qquad \int 3x^2 + 4 \, dx.$$

You could sensibly answer by saying that they're both integrals. You might think to add that the first is a definite integral and the second is an indefinite integral. But we can get more insight by thinking a bit further. If we do the calculations to find the definite integral, we get

$$\int_1^5 3x^2 + 4 \, dx = \left[x^3 + 4x \right]_1^5 = 125 + 20 - 1 - 4 = 140.$$

This is a number. It would be, because a definite integral gives us the area under a graph between the two specified endpoints.[4] So, in an important

[3] You should use the phrase "rational number" rather than "fraction" when that's what you mean. "Fraction" tends to get used more broadly and imprecisely, and it is more mathematically sophisticated to use the proper word.

[4] If we were using this integral to find the area under a graph of velocity against time, this number would tell us about displacement from an original position, but I'll

sense, $\int_1^5 3x^2 + 4 \, dx$ is a number. It doesn't *look* like a number—it looks more like an instruction to do something. But we've seen that things that look like instructions or processes at one level can be thought of as objects at a more sophisticated level (for you, $5 + 3$ was once an instruction to count, but it isn't any more). This definite integral is a fancy way of representing the number 140. Of course, it also carries with it a lot of extra information about the relationship between this number and a particular function.

Now, what about the indefinite integral? If we do the integration here we get

$$\int 3x^2 + 4 \, dx = x^3 + 4x + c.$$

You've probably written this sort of thing so many times that you've stopped thinking about what it means, so let's do that now. Clearly the result of this calculation is not a number. It is not even a single function, because the c represents an arbitrary constant. So, in fact, the result of the calculation is an infinite set of functions: all the things of the form $f(x) = x^3 + 4x + c$. Algebraically this makes sense because, if we differentiate any of these functions, we get back $3x^2 + 4$. Graphically it makes sense because all of these functions are vertical translations of each other, so they have the same gradient everywhere. Considered in terms of object types, however, the difference between this and the definite integral is huge. The integrals look very similar on the page, and indeed we go through the same sort of process to calculate either one of them. But the definite integral is a number and the indefinite integral is an infinite set of functions.

2.4 Objects as the results of procedures

This sort of insight can be very useful. In particular, it is related to the comments in Chapter 1 about students giving the wrong kind of object in response to an exercise. Recall I said that this sometimes seems to happen

assume we're just dealing with pure mathematics here so we can think about numbers alone.

when people try to match a set of symbols in a question with a similar set of symbols somewhere in their notes. The integrals example shows that things that appear very similar on the page can represent dramatically different objects.

To give you a sense of how this can happen in upper-level mathematics, I'll describe a question that a lot of my students got wrong, for exactly this kind of reason. The question was structured like this:

Find $\dim(\ker(\phi))$ where $\phi : \mathbb{R}^4 \to \mathbb{R}^3$ is given by the matrix $\begin{pmatrix} 1 & 0 & 4 & 4 \\ 2 & 2 & 3 & 1 \\ 5 & 2 & 2 & 2 \end{pmatrix}$.

You won't understand this if you haven't yet studied a course called Linear Algebra, so I'll explain just enough for you to see what went wrong for my students, couching my explanation in terms of object types. I'll split this up to make it easier, but it might seem very abstract, so you might want to read it again when you've taken such a course.

- We'll start with ϕ (pronounced "phi"). Here, ϕ is a kind of function known as a *linear transformation*. Instead of taking numbers as inputs and outputs it takes 4-component vectors (things of the form (a, b, c, d)) as inputs and gives 3-component vectors (things of the form (a, b, c)) as outputs. That's what $\phi : \mathbb{R}^4 \to \mathbb{R}^3$ means.
- $\ker(\phi)$ is an abbreviation for the *kernel* of ϕ, which is the set of all the 4-component input vectors for which ϕ gives the output vector $(0, 0, 0)$ (some of the 4-component vectors get sent to $(0, 0, 0)$, some don't). Because of this, $\ker(\phi)$ is a set of 4-component vectors.
- Now, it turns out that we can talk about something called a *basis* for a set of vectors such as $\ker(\phi)$, where a basis is a small number of vectors which we can add together in various combinations to "build" all the others. So a basis for $\ker(\phi)$ is another, possibly smaller, set of 4-component vectors.
- Finally, dim means *dimension*, which is the number of vectors we need to make a basis. The upshot of this is that the answer to the overall question should be a number (read the question again to make sure you can see this).

Most of my students found a basis for ker(ϕ) and stopped there, missing off the final step. This meant that their answers were wrong: they gave a set of vectors when they should have given a single number. A charitable interpretation of what happened is that they forgot to do the last step. This doesn't seem very likely, though—the last step just involves counting how many vectors you have, and the most the answer could possibly be is four (you'll find out why in Linear Algebra). So it's not exactly difficult. In fact, it is easier than most of the other steps. I think that what actually happened is that they didn't really understand the question; instead of thinking about what type of object they were expecting, they found a worked example in their notes that involved a similar symbol layout, and copied that. Unfortunately, the worked example must have been about finding a basis, which is useful for this question but is not enough to fully answer it.

Thus, failing to fully understand can lead to an answer that cannot be right because it is not the right type of object. To mathematicians, such differences are very important. So, as you progress through your major, it is a good idea to adjust your thinking so that you focus less on notational similarities and more on underlying objects and structures.

2.5 Hierarchical organization of objects

Mathematics is not unique in its use of abstract concepts. All kinds of other fields have them: things like "gravity" and "justice" and "anger." No-one's ever seen "a gravity," obviously. But mathematics might be unique, or at least extreme, in the extent to which its abstract objects are organized into hierarchical structures.

We've just seen an illustration of this in the question about dim(ker(ϕ)). Notice that the overall question about dimension doesn't make sense until you know what a basis is and know that a kernel is something that can have a basis. The idea of kernel doesn't make sense until you understand transformations as functions that work with vectors as inputs and outputs. And so on. The concepts could thus be thought of as stacked on top of each other, as illustrated in the diagram on the next page. Understanding the lower ones is necessary in order to make sense of the higher ones.

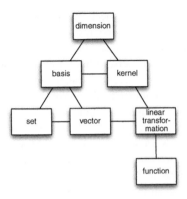

For a more familiar example, consider differentiation. The idea of differentiation doesn't make sense unless you understand about functions, and to make sense of functions you need to know about numbers. Again, these concepts can be thought of as stacked on top of each other, because we can make sense of objects at one level by understanding how they relate to objects at lower levels. This is what I mean by mathematics being hierarchical.

Now, sometimes learning to work at a new level in such a hierarchy is hard, because it involves learning to work with a new type of object. This might involve compressing a process so you can think about it as an object, as in the opening discussions about numbers and functions. Such compression can be difficult because you have to pull yourself up by your own bootstraps: you can't really apply a higher-level process until you can think about the existing thing as an object, but there's no real need to think about the existing thing as an object until you want to use it as an input for a higher-level process. As a result, when learning to work at a higher level, you might go through a phase of being able to "push the symbols around" and get correct answers, without really understanding why what you're doing makes sense. This phase will probably pass if you persevere, but you might be able to understand more quickly if you can identify the source of your confusion as the introduction of a new object type.

That said, I do not mean to argue that there is anything inherently wrong with pushing symbols around. One source of power in mathematics is our ability to do exactly that. To reiterate an earlier point, I don't need to gather 6015 oranges and another 5 oranges to work out what $6015 + 5$ is—I can use the regularities of the symbol system and just get on with manipulating. It's extremely useful to be able to "forget what it's all about" and manipulate symbols according to the standard rules, and I will discuss this further in Chapters 4, 5, and 6. As I've explained above, though, if symbol-pushing is all you can do, then you're vulnerable to certain kinds of error. So it's often worth aiming to develop an object-based understanding as well.

2.6 Turning processes into objects

In undergraduate mathematics you will encounter many new processes, and sometimes you will be expected to start treating these as objects very quickly. For instance, if you study something called Abstract Algebra or Group Theory, you will come across the idea of a *coset*. You will first experience cosets via a process—you'll be shown how to calculate cosets of certain elements in various groups. But, probably in the very next lecture, your professor will treat these cosets as objects. In particular, he or she might perform operations on these cosets—add them together or whatever. If you're still associating the word "coset" with the calculation process, you'll be flummoxed, in exactly the same way that 8 year-olds would be flummoxed about adding 19 and 27 if they were still trying to count everything. Thus, it's a good idea to be ready for these process-to-object transitions.

To illustrate how, here is something that might help you to better understand the differentiation example. At some point, you will study *differential equations* (you might have started already if you are taking the later courses in a Calculus sequence). To understand what differential equations are, we can compare them with ordinary algebraic equations. When solving an ordinary equation, we use x to stand for a solution while we do some algebra, and we expect the solutions to be numbers.[5]

Differential equations are different. Instead of relating an unknown number x to various powers of itself (for instance), they relate an unknown function $y = f(x)$ to various derivatives of itself (and sometimes to other functions of x). So a differential equation might look like this:

$$\frac{d^2 y}{dx^2} + 6x \frac{dy}{dx} - y = 0.$$

The solution to such an equation will be a *function*: any function $y = f(x)$ that satisfies the differential equation. This makes sense because y is a function of x, and so is dy/dx, and so is $d^2 y/dx^2$. So all we're really doing is saying that adding together some combination of these three functions gives zero. If you're not aware of it in this way, you can end up learning a lot of procedures for solving differential equations without really understanding what these achieve. Your understanding will probably be better if you remember why the solution to a differential equation should be a function.

2.7 New objects: relations and binary operations

In cases where learning new mathematics involves treating existing processes as objects, the language often helps us out. Teachers go around saying "**the function** sine x," for instance. However, in other cases we create new objects in other ways, and sometimes the language is not so helpful, because the things we treat as objects can initially seem too big or too spread out to be thought of as "things" in a meaningful way.

An excellent example of this is the notion of "=" ("equals"). Young children, in fact, tend to read "=" as an instruction to do something. They see $5 + 8 =?$ and treat the "=" as though it can be read as "calculate the answer, please." This might not seem problematic, but it really can be. Children with this conception of what "=" means tend to get very confused

---

[5] Maybe the manipulations will lead to the conclusion that, in fact, there are no solutions, but we usually proceed as though we'll find some in the meantime.

by problems like $6 + ? = 20$, because the "instruction" is in the wrong place (they might give the answer 26—can you see why?). They get even more confused by statements like $6 + 14 = 14 + 6$, because in such cases there appears to be "nothing to do" at all.

Successful students learn that "=" just means that the stuff on one side is exactly the same as the stuff on the other side, although these might be expressed in different forms. Nonetheless, it might surprise you to learn that mathematicians treat "=" as a mathematical object in its own right. Specifically, it is an example of a *relation*. Now, there might seem to be no need to give it such a name. Saying "equals is a relation" doesn't seem to give us any more power in solving equations, for instance—you've been doing that for years without this knowledge. So why would mathematicians do it? Well, considering "=" as an object allows us to ask how it is similar to, and different from, *other* relations. Just as 5 is similar to 2 in that it is prime but different in that it is odd, we can compare "=" with other relations, such as "<". Notice, for instance, that while "=" is meaningful for both numbers and functions, it's really not clear that "<" is. What would it mean to say that $x^2 < \sin x$? We could choose to *define* "less than" for functions in a certain way, but how to do so isn't obvious. That's one difference: some relations work naturally on various types of object and some don't. Another difference is that the relation "=" is what we call *symmetric*. We know that if $a = b$ then we also have $b = a$. This certainly isn't the case for "<", even when it is defined. For numbers, it is never true that if $a < b$ then $b < a$.

The important thing to notice here is that we are talking about "=" and "<" as though they were objects, and asking about their properties: "=" is symmetric, "<" is not; "=" is well-defined if we are talking about functions, "<" is not.

The upshot of this is that at some point, in a course called Foundations or Mathematical Thinking or Introduction to Reasoning, your professor will write something like "For every a and b, if $a = b$ then $b = a$," and your instinct will be to wonder why this person is saying something you've known for years as though it's supposed to be new to you. That's not their intention. What they're doing is treating something you've known about for years as an object in its own right, so that they can compare its properties with those of other objects of the same type (and perhaps go on to prove some general theorems about objects that all have some property in common). It takes a bit of intellectual discipline to deal with this. For a while your brain won't want to let you think of "=" as an object because it just seems too weird. But once you get the hang of it you will see this kind of thing happening all over mathematics.

To illustrate further, I'll give another example: a type of object called a *binary operation*. Addition is an example of a binary operation. It takes two objects (hence "binary") and does something with them (hence "operation") to give another object of the same kind. Thinking of "+" as a binary operation means thinking of it as an object in its own right. Again, we can see why this might be useful if we compare it with other binary operations. Some other long-familiar binary operations are "−", "×", and "÷". In what ways are these similar to and different from the binary operation "+"? One way is that "+" is *commutative*, which means that for any two numbers a and b, we always have $a + b = b + a$. Is that true for "−"? No, certainly not. For instance, $7 - 3$ is not the same as $3 - 7$. So subtraction is not commutative. Is multiplication commutative? How about division?

Here, incidentally, is another opportunity to notice that things that look similar on the page might represent dramatically different kinds of object. We can write $3 < 7$ and $3 - 7$, and they don't look that different—each has two numbers with a symbol between them. But the symbol in the first case represents a relation, and the whole expression is a statement about a relationship between the two objects. The symbol in the second case represents a binary operation, and the whole expression is a number. That's not the same kind of thing at all.

2.8 New objects: symmetries

New types of object can be geometric in origin as well as algebraic, and one such type of object is a *symmetry*. You probably first met the idea of symmetry in a lesson that involved turning a mirror around on some diagrams and noticing when looking in the mirror seemed to give back the original. At some point, you probably also turned the diagrams themselves around and noticed that, say, turning through 90° gave you back a diagram that looked the same as the original. At that point, you had to actually *do* something, interact with a particular diagram in some way, to think about symmetry.

These days, you can talk about symmetry as a property. You can say things like "A square has four lines of reflection symmetry," and "Figure 2 has rotational symmetry of order 3." In these sentences, you're treating the diagram (or the imaginary square) as an object, and you're giving information about its properties as these relate to various types of symmetry. Notice, though, that the language has begun to change a bit here. We can talk about *types of symmetry*, as though symmetry comes in various kinds, like cheese. That makes it sound at least like a type of stuff (aliens wouldn't be able to tell just from the language which of *types of cheese* and *types of symmetry* refer to real, physical objects). Mathematicians go one step further, though, and treat individual "symmetries" as mathematical objects in their own right.

To understand this it is probably a good idea to start by thinking about rotational symmetry. More specifically, to think about rotations. At this point the language really starts to help. You can say things like "**a rotation** through 60 degrees about the point $(0, 0)$" and you certainly know what that means. You're also probably okay with "a translation" and "a reflection." For the latter, you have to imagine turning a piece of paper over rather than just sliding it around, but that's okay. You can probably also imagine combining rotations by doing one after the other. Indeed, you can probably imagine combining a rotation with a reflection, or a rotation with a translation, or a translation with a reflection.

Of course, it only makes sense to refer to these motions as *symmetries* if they give us back the initial configuration. For instance, if we imagine an infinite plane tiled with regular hexagons, then one symmetry will be a rotation through 120° about a point where three hexagons meet. Another symmetry will be a translation through a distance that is three times the length of one side of a hexagon, in a direction parallel to that side. Either of these motions gives back the initial configuration and, further, if we "perform" these symmetries one after another, we get another one. So we're not just treating symmetries as objects, we're treating them as objects that can be combined via a higher-level process that could be expressed informally as something like "do-one-then-the-other." Formally, this combining is called *composition*, which is not a coincidence: it is just like composing functions, because a symmetry of this kind can also be thought of as a function that takes every point on the plane and sends it to another point. Composition, you might notice, amounts to a binary operation on symmetries—if we compose two symmetries, we get another one. Is this binary operation commutative?

To see how mathematicians use these ideas more extensively, consider the symmetries of a simpler configuration: a single equilateral triangle. If we want to talk about the triangle's symmetries, it helps to give them names. Say we use R to denote a rotation clockwise through 120° about the centre of the triangle. We might as well then use R^2 to denote a rotation clockwise through 240°, because it amounts to the same thing as doing R twice ("composing R with itself"). One thing to notice is that if we perform R then perform R^2 we get back to where we started, so it would also be handy to have a way of representing "where we started" or "do nothing." Mathematicians usually call this *id* or *I* (short for the *identity transformation*). Then we have the reflections. If we label the triangle so the top vertex is number 1 and then go around clockwise numbering the others 2 and 3, we could use r_1 to denote reflection in the line through vertex 1, and so on.

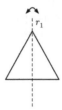

Once we've sorted that out we can compile a list of the outputs of all possible combinations of symmetries. For example, if we do R then r_1, we end up in the same configuration as if we'd just done r_2 (check—cutting a triangle out of paper might help). If we want, we can record all these combinations in a table:

	id	R	R^2	r_1	r_2	r_3
id	id	R	R^2	r_1	r_2	r_3
R	R	R^2	id	r_3	r_1	r_2
R^2	R^2	id	R	r_2	r_3	r_1
r_1	r_1	r_2	r_3	id	R	R^2
r_2	r_2	r_3	r_1	R^2	id	R
r_3	r_3	r_1	r_2	R	R^2	id

Having done that, we can stop thinking about the triangle and just think about the ways in which the symmetries interact with each other. This is similar to what happens when we stop thinking about sets of five oranges and just think about how the number 5 interacts with other numbers. In essence, what we've done is construct *a new kind of arithmetic*. The objects in this new arithmetic are symmetries of the triangle, combined via the operation of composition. Notice that this new kind of arithmetic is, in some sense, "smaller" than ordinary arithmetic. Instead of involving infinitely many objects (numbers) and several operations (addition, subtraction, etc.), we have just six objects (symmetries) and one operation (composition).

This type of thinking happens a lot in Abstract Algebra and Group Theory, but also in other areas of mathematics. We think about a type of object, we think about how we might compare such objects in terms of their properties, and we think about ways of combining them to get new objects of the same kind. If you can keep that in mind, you'll probably find that Abstract Algebra isn't as abstract as it first appears. You'll also find that you've gained some power in thinking about mathematical statements of various types. We'll consider how in the next chapter.

SUMMARY

- Mathematical concepts such as numbers and functions are often first encountered via actions involving physical objects. Later, people learn to think of these concepts as processes, and eventually as objects.
- Thinking of numbers in terms of the way they interact with other numbers and operations is essential because counting takes time.
- It is possible to understand differentiation as a higher-level process that operates on functions, which helps in understanding differential equations.
- Symbolic expressions that look similar on a page might represent very different types of objects. These distinctions are important to mathematicians.
- Students who do not fully understand a question sometimes give an answer that cannot possibly be right because it is not even the right type of object.
- Mathematical objects can be thought of hierarchically organized.

- Learning about objects at a higher level can be difficult when it involves thinking of a process as an object in its own right; in upper-level courses, you will sometimes find that new processes get treated as new objects very rapidly.
- It might take some intellectual discipline to learn to think of concepts such as relations, binary operations, and symmetries as objects, but it makes sense to do so because we can compare their properties.
- If we have a set of objects any two of which can be combined to give another of the same kind, we have a way to construct a new arithmetic.

FURTHER READING

For an accessible introduction to the ways in which mathematicians think about abstract concepts in advanced mathematics, try:

- Gowers, T. (2002). *Mathematics: A Very Short Introduction*. Oxford: Oxford University Press.
- Katz, B. P. & Starbird, M. (2012). *Distilling Ideas: An Introduction to Mathematical Thinking*. Mathematical Association of America.
- Stewart, I. (1995). *Concepts of Modern Mathematics*. New York: Dover Publications.

For a more textbook-like introduction to formal ideas about numbers and axiomatic systems, try:

- Krantz, S. G. (2002). *The Elements of Advanced Mathematics (2nd Edition)*. Boca Raton, FL: Chapman & Hall/CRC.
- Liebeck, M. (2011). *A Concise Introduction to Pure Mathematics (3rd Edition)*. Boca Raton, FL: CRC Press.
- Stewart, I. & Tall, D. (1977). *The Foundations of Mathematics*. Oxford: Oxford University Press.

For more on childrens' understandings of mathematical concepts, try:

- Ryan, J. & Williams, J. (2007). *Children's Mathematics 4–15: Learning from Errors and Misconceptions*. Maidenhead: Open University Press.

CHAPTER 3

Definitions

This chapter is about the place of definitions in mathematical theory. It compares definitions with axioms and theorems, and explains factors that might affect how a definition is formulated. It discusses ways in which mathematical definitions differ from dictionary definitions, and describes strategies for understanding definitions of both familiar and unfamiliar concepts.

3.1 Axioms, definitions, and theorems

In the previous chapter, I argued that we can often think about mathematical concepts as objects in their own right. It can be useful to think in this way when trying to understand mathematical sentences of various kinds, and in this chapter I will explain how this applies in relation to definitions. In the following chapters I will explain how it applies in relation to theorems and proofs.

We'll begin by making sure we know what axioms, definitions, and theorems are, focusing in particular on how they differ (we will deal with proofs in Chapter 5). This is worth a moment, because undergraduate students often can't say what the differences are, even when they have been a mathematics major for a year or more and have written these words dozens of times in their lecture notes. My descriptions will be fairly informal—I'm sure a logician would give a more sophisticated explanation. But they should be enough to focus your attention on the status of each within mathematical theory.

3.2 What are axioms?

An axiom is an assumption that we agree to make, usually because everyone agrees that it is sensible. For instance, one axiom for the real numbers is that for every two real numbers a and b, $a + b = b + a$. As I said in Chapter 2, this is called commutativity (of addition), and this property is not necessarily shared by other binary operations. That's one reason to bother writing it down. Another reason is that mathematicians like to get their assumptions out in the open. This might hardly seem worth it in cases like this, where everyone obviously agrees. But there can be cases in which two people inadvertently assume that different, more subtle axioms hold, and then get confused by each other's claims. To avoid this, it's good mathematical practice to state all our axioms clearly up front.

In the context of this discussion, one important thing to remember is that we do not try to prove axioms. An axiom is something that we all agree to assume is true so that we can get on with using it to prove more interesting things. (I'm simplifying a bit here—see Chapter 5 for further discussion about the relationship between axioms, definitions, and proof.)

3.3 What are definitions?

Unlike an axiom, a definition has nothing to do with something being true. It just tells us what a mathematical word means. A definition might define a type of object, or it might define a property; below you can see some illustrations. Have a go at understanding each one, but don't worry if you don't, because we're just looking at their structure for the time being. Later on we'll look at the meanings in detail.

Definition: A number n is *even* if and only if there exists an integer k such that $n = 2k$.

Definition: A function $f : \mathbb{R} \to \mathbb{R}$ is *increasing* if and only if for every $x_1, x_2 \in \mathbb{R}$ such that $x_1 < x_2$, we have $f(x_1) \leq f(x_2)$.

Definition: A binary operation $*$ on a set S is *commutative* if and only if for all elements s_1 and s_2 in S, $s_1 * s_2 = s_2 * s_1$.

Definition: A set $X \subseteq \mathbb{R}$ is *open* if and only if for every $x \in X$ there exists $d > 0$ such that $(x - d, x + d) \subseteq X$.

Notice that in each definition, just one thing is being defined. The first, for instance, defines *even* as a property of numbers; the second defines *increasing* as a property of functions. Notice also that the hierarchical nature of mathematics is very visible in definitions. In order to understand what *commutative* means, we need to know about sets and binary operations.

Now, essentially, what I've just said is that a definition tells us what a word means, which you already know because you've been using dictionaries for a very long time. But there are two very, very important differences between dictionary definitions and mathematical definitions. If you want to understand advanced mathematics, it is vital that you understand these differences.

The first difference is that when mathematicians state a definition, *they really mean it*. They don't mean that this is a good description of the majority of cases, but that there might be exceptions out there somewhere. This is **not** the way dictionary definitions work. If you took two dictionaries and looked up an everyday concept, either a concrete one (like "table") or an abstract one (like "justice"), would you expect the definitions to be identical? Probably not. Probably, in fact, you would expect to be able to find things in the world that satisfy one definition but not the other. Or to find something that you would want to call a "table" or "justice" but that didn't really satisfy either definition.[1] Or to find something for which, even though there is a definition, people would disagree.

Now, compare this with what would happen if you took a mathematics book and looked up a definition of "even number." Would you expect to be able to find an even number that did not satisfy the definition? Or a non-even number that did, nonetheless, satisfy the definition? Absolutely not. Exceptions simply don't exist. It's not like you could find a number so big that it could be even without being divisible by 2, or something like

---

[1] Is the 12 ft tall "table" that I recently saw in an art exhibition a table? Sort of, but it wouldn't meet functional criteria such as being a flat surface you could rest things on—no-one would be able to reach, and anyway touching the exhibit was not allowed.

that.[2] If you took two textbooks, you might not expect the two definitions to be phrased in exactly the same way, but you would expect them to be logically equivalent. That is, you would expect that all and only the things that satisfy the first definition would also satisfy the second. So, that's one difference: mathematical definitions mean exactly, *exactly* what they say. There are no exceptions, and different phrasings might exist but these must be logically equivalent.

The other difference, which is partly a consequence of the first, is that mathematical definitions are operable in a way that dictionary definitions are not. This means that they contain some information that we can actually manipulate in an algebraic or logical argument. I'll describe what I mean by this later in this chapter.

A final point here is that we do not "prove" definitions. We can't, because there's nothing to prove: definitions just capture conventions in which we all agree to use a word to mean exactly the same thing. A professor might, at some point, explain to you how a definition captures an intuitive idea, but this isn't the same as proving it.

3.4 What are theorems?

That just leaves theorems. Whereas a definition states what we mean by a word describing a mathematical object or property, a theorem tells us about a relationship between two or more types of objects and properties; it assumes that we already know what those objects and properties are. To illustrate, here are some theorems.[3] Some you will understand, and some you won't, but again we're just looking at their structure for now.

Theorem: If *l*, *m*, and *n* are consecutive integers, then the product *lmn* is divisible by 6.

[2] Though I did once hear a story about a PhD student who confused some parents at an open day by telling them very seriously that his research involved searching for a new even number between 2 and 4. Much mathematical humor rests on this kind of thing.

[3] If you already have some experience in upper-level mathematics, you might notice that some of these theorems could be stated more precisely. I have stated them simply here so as not to confuse readers new to advanced mathematics, but I will discuss this issue in Chapters 4 and 8.

Theorem: If f is an even function, then $\dfrac{\mathrm{d}f}{\mathrm{d}x}$ is an odd function.

Theorem: If $ax^2 + bx + c = 0$, then $x = \dfrac{-b \pm \sqrt{b^2 - 4ac}}{2a}$.

Theorem: If $\phi : U \to V$ is a linear transformation, then
$$\dim(\ker(\phi)) + \dim(\mathrm{im}(\phi)) = \dim(U).$$

Hopefully you can see how these differ from definitions. In the second one, for instance, it is assumed that we already know the meaning of *function*, *even*, *odd*, and $\mathrm{d}f/\mathrm{d}x$. None of these is being defined here; instead we are being told about a general relationship between them.

Of course, if you don't know the meanings of all the words and symbols, you will probably feel that you don't understand the theorem. Notice, however, that even if you don't understand a theorem, you should be able to recognize that it has a theorem-like structure. The above theorems are all written in this form:

If this thing is true, *then* this other thing is true as well.

In each case, the bit that goes with the *if* is called the premise or the assumption or the hypothesis (I realize it's a bit annoying having multiple words for the same thing—sorry about that). The bit that goes with the *then* is called the conclusion. It's actually not quite this simple in real life, because there are a few ways of phrasing theorems that don't make the "if…then…" structure so obvious. I'll discuss that further in Chapter 4.

In one sense, theorems are similar to axioms: they tell us true things about relationships between concepts. The difference is that we go to the trouble of proving theorems. Sometimes we do this because theorems are far from obvious (do all of the above theorems seem obviously true to you?). Sometimes, though, theorems are pretty obviously true, and we do it for the more subtle reason that proving them allows us to see how they all fit together to form a coherent theory. I'll have more to say about this in Chapter 5.

One thing to notice before we move on, however, is that on a surface level, axioms, definitions, and theorems all look similar. Indeed, I once had a student tell me that he thought definitions and theorems were "kind of the same" for this reason. You will be able to tell which is which in your lecture notes because the axioms will be labeled "Axiom," the definitions will be labeled "Definition," and the theorems will be labeled "Theorem" (seriously).[4] But they're often of similar lengths and they usually involve a combination of words and symbols, so seeing one on a page would not tell you about its logical status within a mathematical theory. Hopefully you can see that the logical structures of axioms, definitions, and theorems differ in important ways, and that they each have a different status. Bear this in mind and you'll be better able to keep track of the global structures of your mathematics courses.

3.5 Understanding definitions: even numbers

The remaining chapters in this part of the book will address ways of interacting with definitions, theorems, and proofs. In this chapter, we will focus on understanding definitions by thinking in terms of abstract objects. We'll consider a very simple definition to start off with.

Definition: A number is *even* if it is divisible by 2.

"Well, yes," you'll be thinking, "obviously." In fact, though, this is probably not how your professor will write this definition. You're more likely to see something like this (as it was presented above):

Definition: A number n is *even* if and only if there exists an integer k such that $n = 2k$.

You could be forgiven for thinking that this over-complicates things. But it has some advantages. The first is precision. We can see this by comparing with the kind of thing that students sometimes write when they try to define *even*. They tend to say things like "Even is when it's divisible by 2." Clearly that captures the key idea, but it isn't very precise. For a start,

---

[4] This might not be quite true all the time in every course—sometimes professors introduce ideas more in narrative paragraphs—but you will find that it's almost always the case.

what is "it?" Clearly "it" is supposed to be a number, but this number isn't introduced properly. In contrast, the better definition tells us explicitly that we are dealing with a number and gives it a name, n. For another thing, the student definition contains the locution "is when." This tends to sound clunky, in mathematics but also in other fields. In mathematics, if you find yourself writing "it" or "is when," you should probably consider rephrasing (for lots of information on improving your mathematical writing, see Chapter 8).

The second advantage is the one mentioned earlier: operability. We can operate with mathematical definitions to prove things. The better definition gives us a way of capturing and manipulating even numbers because it states what divisibility means in algebraic terms. We could, for instance, use it to prove that any integer multiple of an even number must also be even, perhaps by writing something this:

Claim: Any integer multiple of an even number is also even.

Proof: Suppose that n is an even number.

Then (by definition) there exists an integer k such that $n = 2k$.

Now, let z be any integer.

Then $zn = z(2k)$.

But we can rewrite this as $zn = 2(zk)$.

Now zk is an integer, so zn is even because it can be written in the required form.

We could have the same ideas if we were working with the imprecise student definition. But the better version gives us a leg-up for writing arguments like this by providing some notation. I'll discuss this further in Chapters 5, 6, and 8 (Chapter 5 also has information on why we would bother proving such an obvious thing).

Returning to our discussion of the definition in itself, it is not difficult to think about how it applies to particular mathematical objects. The whole definition is "about" integers—whole numbers. We can easily check that the definition corresponds to an informal understanding of evenness, by thinking about how it applies to some of these numbers. For the even number 24, the k is 12. For the even number 0, the k is 0. For the even number 156 206 749 386, I can't be bothered to work out what the k is, but I'm convinced that I could do it if I had to—an appropriate k certainly exists.

By thinking about even numbers, though, we've only done half of the job. We expect the definition to hold for every even number, and it does. But, just as importantly, we expect it to *exclude* all the numbers that are not even. This, too, is easy to think about in this case. For the number 3, there is no appropriate integer k. For the number -10.2, there is no appropriate integer k, and so on. Thus we can satisfy ourselves that we understand the definition, and that it captures the notion of evenness in a reasonable way.

3.6 Understanding definitions: increasing functions

The previous illustration was straightforward because integers are very familiar. We will now think about a case in which the idea is, again, familiar and natural, but the definition is a bit more complicated. We'll use the definition of *increasing* for functions.

Probably you have a pretty clear idea of what it ought to mean for a function to be increasing. It's likely that the idea brings to mind a graph, perhaps one that looks like this:

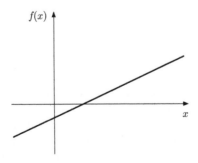

With that in mind, let's look at the definition.

Definition: A function $f : \mathbb{R} \to \mathbb{R}$ is *increasing* if and only if for every $x_1, x_2 \in \mathbb{R}$ such that $x_1 < x_2$, we have $f(x_1) \le f(x_2)$.

It's a good idea to work out how this relates to the graph. First, the definition talks about a function f that maps real numbers to real numbers (that's what $f : \mathbb{R} \to \mathbb{R}$ means). That's what Cartesian graphs are designed to show: for any given (real number) input on the x-axis, they allow us to read a (real number) output on the y-axis. Next, the definition introduces two real numbers and calls them x_1 and x_2 ($x_1, x_2 \in \mathbb{R}$ means that x_1 and x_2 are elements of $\mathbb{R}$). We'd like to put these on the diagram, but first we have to work out where. We have two sensible candidates for this: they could be inputs (on the x-axis) or outputs (on the y-axis). When mathematicians name things, the names are usually sensible, and here we have things called $x_{\text{something}}$, so it's a good bet that the place for them is on the x-axis. We can confirm this by looking at the rest of the definition, which involves taking f of each one. The only other information we're given is that we're considering cases where $x_1 < x_2$, so let's put them in that configuration, and label their f-values:

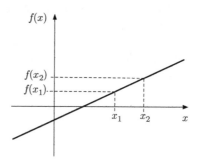

Now, the definition says that whenever we have $x_1 < x_2$, we should have $f(x_1) \le f(x_2)$. That certainly seems to be the case in the picture. Technically, though, we've only thought about one pair of numbers x_1 and x_2, and the definition says something about every such pair. But graphical representations are useful for thinking about general ideas: we can imagine x_1 and x_2 sliding around on the x-axis (keeping $x_1 < x_2$, of course) and think about what would happen to $f(x_1)$ and $f(x_2)$. Doing this should convince you that the definition is in accord with our intuition for this particular function. What would you do to convince yourself that it is in accord with our intuition for increasing functions in general? Yep— imagine the graph of the function changing, too.

So, we have checked that the definition seems to hold for functions that we think should be classified as increasing. As before, though, we should also check that it does not hold for functions that we think should not be classified as increasing. For instance, the definition does not hold for the function shown below, because we have $x_1 < x_2$ but we don't have $f(x_1) \leq f(x_2)$. Indeed, if we imagine sliding x_1 and x_2 around, we can see that we never have $f(x_1) \leq f(x_2)$.

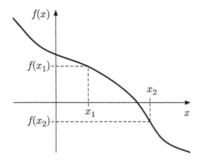

We should also make some effort to think about more complicated cases. Consider, for example, the function represented in the diagram below.

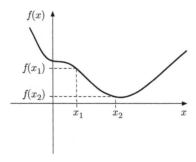

Here we have $x_1 < x_2$ but we don't have $f(x_1) \leq f(x_2)$. However, in this case, if we imagine sliding x_1 and x_2 around, keeping $x_1 < x_2$, we find that in some places we *do* have $f(x_1) \leq f(x_2)$. What does that mean, in terms of our definition? Reading it again carefully, we find that it says "for **every** $x_1, x_2 \in \mathbb{R}$ such that $x_1 < x_2$, we have $f(x_1) \leq f(x_2)$." You won't be surprised to learn that when mathematicians say "every," *they really mean it*. In this case, the required relationship does not always hold, so the function does not satisfy the definition, so it is not increasing.[5]

Can you see why it is important to say *not increasing* rather than *decreasing*? In mathematics, opposites don't work the same way as they appear to in everyday conversation. Some functions that are not increasing are, indeed, decreasing. However, some functions that are not increasing have more complicated behaviors. We'll come back to this point in Section 3.10.

3.7 Understanding definitions: commutativity

At this point, we've seen two definitions that apply to familiar objects (numbers and functions, respectively). You'll also meet definitions that are harder to think about because they apply to less familiar objects. Here is an example.

Definition: A binary operation $*$ on a set S is *commutative* if and only if for all elements $s_1, s_2 \in S$, $s_1 * s_2 = s_2 * s_1$.

If you're reading this before taking upper-level courses, then you've only known about binary operations and commutativity since the middle of Chapter 2. You may not have fully understood that information and, even if you did, you have much less to draw on in making sense of this definition. Nonetheless, we can apply the same sort of thinking to it.

First, as with the previous definitions, we can call to mind a particular example of a binary operation and work out how it relates to the

[5] You might find yourself wanting to say that the function is increasing as long as x is to the right of some point on the x-axis. That's fine. Mathematicians sometimes specify that a definition does not hold overall but does hold on some restricted domain.

definition. As before, let's start with one that we know is commutative: addition. Addition is denoted by "$+$", but there is no "$+$" in our definition. However, there is a general name for a binary operation: in this definition it is called "$*$". This is the same kind of notational device as was used in the other definitions. We had a general number called n (which we replaced in our thinking with some particular numbers), and a general function called f (which we replaced in our thinking with some particular functions). Here, we have a general binary operation called "$*$", which (right now) we are replacing in our thinking with the particular binary operation "$+$".

The definition describes this binary operation as being "on a set S." What could that mean? If we look ahead, we see that we're dealing with two elements of S (called s_1 and s_2) and we'll be combining them with the binary operation. When we're dealing with "$+$", we could treat s_1 and s_2 as integers, so that it makes sense to think of S as the set of all integers, $\mathbb{Z}$. Then, as usual, we should establish that the definition applies to our binary operation in the way we expect. In this case it does: for all elements $s_1, s_2 \in \mathbb{Z}$, $s_1 + s_2 = s_2 + s_1$. Then, as before, we can go on to think about more examples. For instance, the binary operation "$\times$" is also commutative on the set $\mathbb{Z}$. So is the binary operation "$+$" on the set of all functions from $\mathbb{R}$ to $\mathbb{R}$ (for any two functions f and g, $f + g = g + f$). As usual, it is also a good idea to think about non-examples. For instance, as we saw in Chapter 2, the binary operation subtraction is not commutative on the integers: there exist pairs of integers s_1 and s_2 such that $s_1 - s_2 \neq s_2 - s_1$. As always, think carefully about how this shows that subtraction on the integers does not satisfy the definition.

For more exotic examples, we can think about newer types of object. We might think about symmetries combined via the binary operation of composition, as in Chapter 2. We didn't introduce any special notation for this binary operation, so we might as well use the general one "$*$". Considering the symmetries of the equilateral triangle, recall that we have $r_1 * R^2 \neq R^2 * r_1$, so composition on the set of symmetries of the equilateral triangle does not satisfy the definition and thus is not commutative. If you know about matrices, you might also know that matrix addition is commutative on the set of all 2×2 matrices, but that matrix multiplication is not. Again, check to see exactly how this knowledge fits with the definition.

3.8 Understanding definitions: open sets

We've now looked at understanding definitions for familiar concepts and for concepts that you've more recently encountered. You will also find that your professors introduce definitions for totally new concepts. Here is such a definition, which you might meet in a course called something like Analysis or Metric Spaces or Topology:

Definition: A set $X \subseteq \mathbb{R}$ is *open* if and only if for every $x \in X$ there exists a number $d > 0$ such that $(x - d, x + d) \subseteq X$.

The word being defined here is *open*. Most people know what *open* means in relation to doors or shops or tins or envelopes or restaurants or cardboard boxes. Unfortunately, none of these provides much intuition for what this word ought to mean when applied to subsets of the real numbers. Not to worry, though, because we can still understand this new meaning by using the strategies we worked with above.

First, let's select an example to think with. We need a subset of the reals (recall that's what $X \subseteq \mathbb{R}$ means), and there is no sense in making life unnecessarily complicated at this stage, so I will pick a nice simple one: the set containing 2, 5, and all the numbers in between. Mathematicians denote this by $[2, 5]$ (the square brackets mean that the endpoints are included). If we want, we can represent this using a number line:

In the diagram on the left, the included endpoints are represented using colored-in dots. These are a bit misleading, of course, because they make the endpoints look "big" when they're not. In the diagram on the right, the inclusion of the endpoints is represented by using square brackets as in the algebraic notation. This is not quite so pretty, and it still doesn't cope with the points being infinitely small. But human beings are good at thinking about diagrams as representing idealized situations, so either will be fine.

Now we can ask whether or not the set $X = [2, 5]$ is open, by ascertaining whether or not it satisfies the definition. The definition says that for every $x \in X$, something is true. You could try to think about all the x's in X at once, but I'd advise against that. Usually it's easier to start by thinking about one of them and then work out whether and how that thinking generalizes. So let's pick an x and label it:

Notice that in picking this x I've avoided the endpoints, and I've chosen a point that's off-center. I've done this because I'm trying to start with an x that has no special properties, in order to maximize the chances that my thinking will generalize to most of the other points in X. I could, if I wanted to, think of the x as a particular number (4, say). But let's see if we can get away without being quite that specific.

The rest of the definition says "there exists a number $d > 0$ such that $(x - d, x + d) \subseteq X$." First, clearly we need to know what $(x - d, x + d)$ means. It's very similar to the notation $[2, 5]$; $(x - d, x + d)$ means all the numbers between $x - d$ and $x + d$ *without* the endpoints $x - d$ and $x + d$. On a number line, that could be represented as below. Notice that there is some leeway in how to represent distances, points, and so on.

Second, we'd like to work out what this d could be in relation to the question of whether $[2, 5]$ is open. We know d is a number greater than 0, but that's all. Again, though, this doesn't matter—we can pick some random-looking d and think about whether or not it does what we want. If I choose a d, I can label $x - d$ and $x + d$ as on the diagram below.

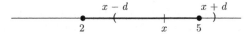

For this d, there's a bit of $(x - d, x + d)$ on the right that goes outside X. Can we conclude, though, that the definition is not satisfied? The answer is "no," because this definition just says "**there exists** a number $d > 0$ such that $(x - d, x + d) \subseteq X$." It doesn't say that this has to work for every possible d, just that there has to be (at least) one. And there is: if we choose a d that is a bit smaller, we can make sure that $(x - d, x + d) \subseteq X$:

Okay, good. That was just for one value of x, though, and our definition says "for **every** $x \in X$ there exists a number $d > 0$ such that $(x - d, x + d) \subseteq X$." However, this thinking does generalize to all the points that are "like" our initial x in the sense that they are in the interior of the interval (that is, not at the endpoints). Even if we take an x that is really, really close to an endpoint, we can pick a d that is small enough to ensure that $(x - d, x + d) \subseteq X$.

That just leaves us with the endpoints themselves. If we have $x = 2$, for instance, can we select an appropriate d? The answer in this case is no. No matter how small we make d, the "left-hand half" of $(x - d, x + d)$ will always[6] be outside X. A similar argument applies at the other end of the interval. So there are points $x \in X$ for which there does not exist an appropriate d. So the definition is not satisfied, and the set $X = [2, 5]$ is not open.

The same kind of thinking applies to any set of the form $[a, b]$. Indeed, we could have just taken a set called $[a, b]$ in the first place, and drawn our first diagram like one of these:

This is an advantage of visual representations—we can often use them to support generalization.

However, we haven't actually found an example of an open set yet. But we can use what we just learned to construct one. It was only the endpoints of the interval that gave us any bother; at all the interior points, we could do what the definition required. So we can construct an open set by just

---

[6] If you are tempted to say "choose $d = 0$," remember that this is forbidden by the definition, which specifies that we must have $d > 0$.

taking the endpoints off. If we take a set X of the form (a, b), then for every $x \in X$ there does exist a number $d > 0$ such that $(x - d, x + d) \subseteq X$. So any set of the form (a, b) is open.

You might like to know that a set of the form (a, b) is called an *open interval* and a set of the form $[a, b]$ is called a *closed interval*.[7] You might also like to know that when we are talking about mathematical sets, *open* and *closed* are not opposites in the everyday sense, just as when we are talking about mathematical functions, *increasing* and *decreasing* are not opposites in the everyday sense. In fact, we can construct sets that are both open and closed, and sets that are neither open nor closed. Be ready for this kind of thing when you are studying advanced mathematics.

This definition was harder to work with than the others, which is what we'd expect because we didn't know what *open* meant when we started. When we have no prior knowledge of relevant examples, we have to work in a more exploratory way, picking an object to think about and asking whether it satisfies the definition. In doing so, we might have to make a lot of choices (which set to pick, which number to pick within it, how specific to go in choosing either one, and so on). This might make you wonder how you should go about choosing example objects. Perhaps, for instance, you'd have been more likely to choose a set X that just contained a finite number of points, like $\{0, 1, 2\}$. What I'd say is: that's fine. When faced with such situations, it's better to pick an object and think about it than to sit there paralysed with doubt about how to pick a "good" one. If the one you choose has the property you're looking for, great. If it doesn't, also great. In the latter case, you've learned something about objects to which the new definition does not apply, and you might also have learned something that will allow you to construct one to which it does. You have to be willing to get your hands dirty when learning new mathematics; to try out lots of things that might help. As you build up a bank of example objects of various types, you'll get better at using them to help you understand new definitions.

---

[7] In fact, the notions of open and closed can be reformulated so as to apply much more generally, beyond just subsets of the real numbers. Can you imagine how open sets might work if we think about areas in a plane, for instance, rather than intervals on a line?

3.9 Understanding definitions: limits

We've now looked at definitions of simple, familiar concepts, and definitions that are harder to think about because they involve concepts that you've met more recently or, perhaps, not at all. You might also come across definitions for which you have a pretty good existing understanding of the relevant objects, but for which the definition is sufficiently complicated that it's hard to relate the two. The classic example is the definition of limit:

Definition: $\lim_{x \to a} f(x) = L$ if and only if $\forall \varepsilon > 0 \; \exists \delta > 0$ such that if $0 < |x - a| < \delta$ then $|f(x) - L| < \varepsilon$.

When I first saw this I thought it was great—it looks like hieroglyphics and is just the sort of thing that easily impresses people who don't study mathematics. But it took me a long time to understand how it captures the notion of a limit, and I've never met a student who finds it easy. I won't give an explanation here as that would constitute a big interruption, but the limit concept is very important and you should persevere when this definition is introduced.

Your professor will explain it, of course. In practice, what usually happens when a professor introduces a new definition is that they then give a couple of examples of objects for which it holds, and perhaps an example of an object for which it doesn't. They might explain in detail, showing exactly how the definition applies in each case, or they might not. Either way, they will expect you to think beyond this very small number of examples; to consider other objects that do and don't satisfy the definition, and to make sure you understand why. *They probably won't tell you that they expect you to do this*. This kind of thinking is so natural to mathematicians that they tend to forget that it's not obvious to everyone. If you want to be a good mathematician, it's a useful habit to develop.

3.10 Definitions and intuition

In the sections above I've discussed ways in which definitions might correspond with your existing, intuitive understandings of various concepts. But it should be clear that intuition is a somewhat personal thing.

While your intuition might coincide with someone else's pretty well in a lot of cases, there's no guarantee that it always will. This turns out to be very important for understanding how to interact with mathematical definitions, and in this section I'll explain how.

To begin, consider the function below.[8]

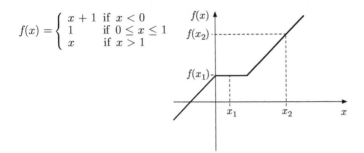

$$f(x) = \begin{cases} x+1 & \text{if } x < 0 \\ 1 & \text{if } 0 \le x \le 1 \\ x & \text{if } x > 1 \end{cases}$$

Intuitively, would you say that this function is increasing, or not increasing? Whatever you think, can you see that someone else might disagree with you?

As mathematicians, we do not have to worry about this sort of uncertainty because we make such decisions by looking at the definition. Here it is again:

Definition: A function $f : \mathbb{R} \to \mathbb{R}$ is *increasing* if and only if for every $x_1, x_2 \in \mathbb{R}$ such that $x_1 < x_2$, we have $f(x_1) \le f(x_2)$.

This function is increasing because, anywhere we put $x_1, x_2 \in \mathbb{R}$ such that $x_1 < x_2$, we have $f(x_1) \le f(x_2)$. Notice that the *or equal to* in the *less than or equal to* is important here. In this case we do sometimes have $f(x_1) = f(x_2)$. This didn't happen with our earlier functions, and it highlights the importance of thinking about how an example might vary.

[8] You might be used to functions defined by a single formula, but this is a perfectly good function that just happens to be defined *piecewise*.

Depending on your initial answer, you might not like the fact that this function is mathematically classified as increasing. You might have been inclined to say that it is not increasing "because of the constant bit," and you might feel uncomfortable with the idea that your intuitive response was, in some sense, wrong. You might, in fact, be inclined to think that the definition must be wrong. If so, try to sit with that feeling without being disturbed by it, because that's what this section is about.

On the other hand, perhaps that example didn't disturb you at all. If that's the case, how about this one?

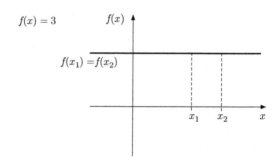

$f(x) = 3$

This function is increasing too. For every $x_1, x_2 \in \mathbb{R}$ such that $x_1 < x_2$, we have $f(x_1) \leq f(x_2)$, as required. We happen to have $f(x_1) = f(x_2)$ everywhere, but that's fine; the definition allows that possibility. In fact, as you might now be anticipating, this constant function is mathematically classified as *decreasing* too, for the same sort of reason (the definition of decreasing is exactly what you'd think it would be). Most people are really disturbed by this. It is far enough from their intuitive response to make them think that mathematicians must have collectively gone mad. This is not the case, however, and I'll explain why now, giving a couple of different reasons.

The first reason is that, when people use a word intuitively, there tend to be small differences in how they do so. You might be inclined to classify a particular function as increasing, but someone else might disagree. This means that the set of functions that you would classify as increasing might overlap with someone else's like this:

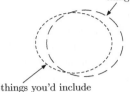
things someone else would include

things you'd include

You and this other person agree about the "central" or "obvious" examples, but disagree about examples near the "boundaries." Now, mathematicians don't like this sort of situation. They abhor it, in fact. They like to be certain about the meaning of each mathematical word, so that they can be sure that everyone is talking about the same objects when they use it. That's why they impose a definition: to decide on a precise way of capturing the meaning, and to agree that they'll all abide by it. When doing so, they try to ensure that the definition captures all the cases for which everyone agrees. However, they also value simplicity and elegance. So, sometimes, they end up with a definition that is actually a little bit different from everyone's natural response. Maybe the defined set overlaps with the ideas of our two people like this:

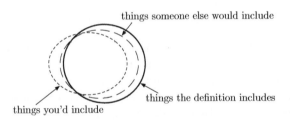
things someone else would include

things the definition includes

things you'd include

This explains what has happened with the definition of *increasing*. We wouldn't want to say "A function is increasing if and only if it is generally going up from left to right," because that wouldn't be very precise. We also wouldn't want to say "A function is increasing if and only if for every $x_1, x_2 \in \mathbb{R}$ such that $x_1 < x_2$, we have $f(x_1) \leq f(x_2)$, unless we have $f(x_1) = f(x_2)$ for every $x_1, x_2 \in \mathbb{R}$, in which case it's just constant," because that wouldn't be very elegant.

In fact, historical decisions about how to formulate a definition can be more subtle than this, which brings us to the second reason why definitions are sometimes a bit strange. Mathematicians aren't usually trying to decide on the precise meaning of a word just for the sake of it. They're usually trying because they want to use it in formulating a theorem and an associated proof. Thus the formulation of the definition might depend on what is convenient for being able to state a nice, elegant, true theorem. The classic illustration of this involves Euler's formula relating the number of vertices (V), the number of edges (E), and the number of faces (F) for a polyhedron. The formula is $V - E + F = 2$, and you can check that this applies in standard cases like a cube or a tetrahedron or any other polyhedron you can think of (try it). However, it does not apply for a solid like this (with a hole through the middle):

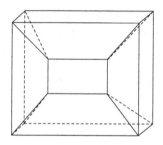

The question, then, is what to do. Do we say "Oh well, never mind," and give up on that lovely elegant formula that applies in so many cases? Or do we say "Right, well then, we'll just define *polyhedron* so that it excludes such cases?" Or do we define a special class of polyhedra for which the formula does apply? Mathematicians don't make the same decision each time they are faced with this sort of choice, but they certainly take such considerations into account.

If you are interested in how mathematicians formulate definitions, you should probably look out for optional courses in things like History of Mathematics or Philosophy of Mathematics. Even if this discussion is a bit too philosophical for your present tastes, however, you need to be aware of the consequences. You need to be aware that you will occasionally come across a mathematical definition that does not quite correspond with your intuitive understanding. It is perfectly okay to find this a bit weird, but you have to deal with it. Mathematicians have made their collective decisions for good, sensible reasons. Those reasons might not be immediately apparent to you but, if the mathematical definition of a term does not match your intuition, it's your intuition that needs fixing up.

I'll end this chapter by saying that if the increasing/decreasing thing is still bothering you, you might like to know that mathematicians also use the following definition:

Definition: A function $f : \mathbb{R} \to \mathbb{R}$ is *strictly increasing* if and only if for every $x_1, x_2 \in \mathbb{R}$ such that $x_1 < x_2$, we have $f(x_1) < f(x_2)$.

Thinking about how that relates to your intuition should restore your faith in mathematicians' good sense.

SUMMARY

- An axiom is an assumption that mathematicians agree to make. A definition specifies the meaning of a mathematical word. A theorem is a statement about a relationship between two or more types of objects and/or properties.
- We do not prove axioms or definitions (though a professor might explain how a definition captures an intuitive idea); we do prove theorems.
- Mathematical definitions are precise, they do not admit exceptions, and they often provide algebraic notation which we can use to construct proofs.
- To understand a definition, it is useful to think about objects to which it does apply and objects to which it does not. Professors will probably expect you to do this, even if they do not say so.
- It is often possible to understand the meaning of a completely unfamiliar definition by constructing examples to which it might apply, then checking.
- Opposites do not work the same way in mathematics as they do in everyday conversation; a function might be both increasing and decreasing, or neither increasing nor decreasing.
- Definitions do not necessarily correspond to intuitive meanings of concepts, especially for "boundary" cases. You should be alert to this; if a definition does not match your intuition, your intuition needs to be amended accordingly.

FURTHER READING

For more on definitions and their place within mathematical theory, try:

- Houston, K. (2009). *How to Think Like a Mathematician*. Cambridge: Cambridge University Press.
- Katz, B. P. & Starbird, M. (2012). *Distilling Ideas: An Introduction to Mathematical Thinking*. Mathematical Association of America.

For a textbook introduction to mathematical concepts from this chapter, try:

- Krantz, S. G. (2002). *The Elements of Advanced Mathematics (2nd Edition)*. Boca Raton, FL: Chapman & Hall/CRC.
- Liebeck, M. (2011). *A Concise Introduction to Pure Mathematics (3rd Edition)*. Boca Raton, FL: CRC Press.
- Stewart, I. & Tall, D. (1977). *The Foundations of Mathematics*. Oxford: Oxford University Press.

For more on working with definitions within proofs, try:

- Solow, D. (2005). *How to Read and Do Proofs*. Hoboken, NJ: John Wiley.

CHAPTER 4

Theorems

This chapter is about understanding both straightforward and complicated theorems. The first half discusses ways to develop understanding by examining examples that satisfy the premises of the theorem, and considering what the conclusion means for these examples. The second half discusses the logical language in which theorems are written, describes differences between everyday and mathematical uses of this language, and highlights things to remember when interpreting theorem statements.

4.1 Theorems and logical necessity

Recall from Chapter 3 that theorems tell us about relationships between mathematical objects and properties. They differ from definitions, because definitions tell us the meaning of just one word. They also differ in a more subtle philosophical sense: where definitions are a matter of convention—mathematicians settle on one agreed meaning, but they could have chosen another—theorems are not. Once definitions are settled, theorems follow by logical necessity. For instance, once we have decided to define even numbers as numbers that are divisible by 2, we must conclude that the sum of any two even numbers is also even. We don't have any choice about that kind of consequence.

In Chapter 3, we looked at understanding definitions by using various different representations of objects. We can apply a similar strategy for theorems, with the additional requirement that we should think about why the statement is true. Below is the list of theorems from Chapter 3. In this chapter, we will concentrate on the first two. We will stop short of proving them, but we will think about how to get ideas that could be helpful in constructing a proof. We will not study the third one; I want to come back to that in Chapter 5. We will, however, have a quick look at the fourth one, to show how we can think in terms of object types even when the concepts are unfamiliar.

Theorem: If l, m, and n are consecutive integers, then the product lmn is divisible by 6.

Theorem: If f is an even function, then $\dfrac{df}{dx}$ is an odd function.

Theorem: If $ax^2 + bx + c = 0$, then $x = \dfrac{-b \pm \sqrt{b^2 - 4ac}}{2a}$.

Theorem: If $\phi : U \to V$ is a linear transformation, then $\dim(\ker(\phi)) + \dim(\mathrm{im}(\phi)) = \dim(U)$.

At the end of the chapter, I will discuss the use of logical language in theorems and in mathematics more broadly. This is important because the way that logical language is used in everyday life is somewhat different from the way that it is used by mathematicians. It is important to be aware of this so that you can correctly interpret mathematical statements, and so that you can make your own mathematical writing accurate.

I should say that although I've called all of our statements *theorems*, there are several words that sometimes get used instead. Some of these are *proposition, lemma, claim, corollary*[1], and the apparently all-encompassing *result*. *Lemma* is usually used for a small theorem that will then be used to prove a bigger, more important one. *Corollary* is used for a result that follows as a fairly immediate consequence of a big theorem. The other terms are more or less interchangeable, I think, though it might be that some professors use them to make subtle distinctions—ask if you're not sure. Such distinctions can be important to a professor who wants to help

---

[1] It is not obvious how to pronounce this. In the UK, we say coROLery, but I once inadvertently had a whole class of American students in fits of laughter over this—they say CORolairy and thought my pronunciation was hilarious.

students discern the structure of a whole course (what's important, what's ancillary). But that's not what I'm doing here so, for simplicity, I'll just call them all theorems.

4.2 A simple theorem about integers

We observed in Chapter 3 that all our theorems have this structure:

If this thing is true, *then* this other thing is true as well.

One important thing to know is that mathematicians interpret such statements as having an implicit "always." For instance, the first theorem should be interpreted to mean that if l, m, and n are consecutive integers, then the product lmn is *always* divisible by 6. The second one should be interpreted to mean that if f is an even function, then df/dx is *always* an odd function. Technically, mathematicians should really write things like this:

Theorem: For all integers l, m, and n, if l, m, and n are consecutive, then the product lmn is divisible by 6.

However, that makes everything a bit longer and, because everyone is familiar with the standard interpretation, people rarely do it.

Once we know this, the first theorem is simple to understand because it is about integers. You probably feel that you understand it without having to do anything, even if you're not sure why it's true. However, there are a couple of points I'd like to make. First, notice that the premise introduces the objects we will be working with, and tells us what we're going to assume about their properties. In this case, the objects are three integers, and the only property is that they are consecutive. The conclusion tells us something that follows logically from the premise. It doesn't tell us *why* it follows, just that it does. Here, the conclusion says that the product of our three consecutive integers will always be divisible by 6.

This means that we can think about this theorem by picking three consecutive integers and multiplying them together. For instance, if we multiply together 4, 5, and 6, we get 120, which is certainly divisible by 6. It was always going to be, though, because 6 was one of the factors we started with. So that wasn't a very illuminating example. How about 7, 8, and 9? If we multiply those together we get 504, and we can easily check that 504 is divisible by 6. We could try other sets of three consecutive numbers too, remembering that $-21, -20, -19$ is a perfectly good set of three consecutive integers, as is $-1, 0, 1$. It is important to consider less obvious cases like these when you want to develop a proper understanding of a theorem.

Now, I'm not advocating that you should spend ages multiplying together triples of consecutive integers. Doing the same calculation over and over again probably wouldn't lead to any insight into why the theorem is true, and certainly would get boring. It might be more useful to consider alternative ways of representing the objects. In this case, we could think about consecutive integers on a number line:

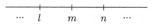

Alternative representations sometimes highlight different aspects of a mathematical situation. Does this one help you notice anything that might be useful? We'll come back to this in Chapter 5, when we look at a proof.

4.3 A theorem about functions and derivatives

For now, we will move on to the second theorem:

Theorem: If f is an even function, then $\mathrm{d}f/\mathrm{d}x$ is an odd function.

Here there is more work to do, because the objects and properties are more abstract. Again, however, the premise introduces the objects we're working with and tells us what properties they have. Here it introduces just one object, a function called f, and tells us that it is even. *Even* is a mathematical property so it has a definition, and if we look it up we find this:

Definition: A function f is *even* if and only if for every $x, f(-x) = f(x)$.

I should stress that you *must* look up definitions whenever you don't know or can't remember what something means. If you don't—if you allow yourself to proceed with only a vague and woolly understanding of a concept—you are very unlikely to develop a good understanding of the theorem.

In this case, our function f has the property[2] that for every x, $f(-x) = f(x)$. Probably you can think of some functions with this property, even if you weren't previously familiar with it. For instance, f given by $f(x) = x^2$ has it. So does f given by $f(x) = \cos x$. From these examples we can easily build more: $f(x) = x^2 + 1$, $f(x) = 3\cos x$, and so on.

An alternative representation is useful here, because we are accustomed to thinking about graphs. If you have studied *even functions* before, you might know that an even function is symmetric about the y-axis. If you didn't know that, you can probably see why it is true by looking at the diagram below and thinking about all the possible values of x. You could also think about similar diagrams for the particular functions we just listed.

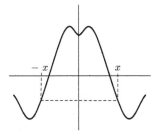

Thinking about examples and graphs can thus help us to understand the premise. But what about the conclusion? The conclusion here says that df/dx is an odd function. First, we can make sure that we know what objects we now have. From Chapter 2, you should be comfortable with the idea that when f is a function, so is df/dx. Perhaps, though, we need to look up the definition of *odd*:

Definition: A function f is *odd* if and only if for every x, $f(-x) = -f(x)$.

---

[2] A definition might also explicitly specify that $(-x)$ must be in the domain of f. In upper-level mathematics, you will find that mathematicians are increasingly careful about technical requirements like this.

That looks very, very similar to the definition for *even*, so make sure you can see where the difference is. Don't be blasé about this kind of thing—sometimes shifting one symbol makes all the difference.

As before, we can check that the conclusion seems to hold for some of the objects that satisfy the premise. Suppose we take $f(x) = x^2$. Then $df/dx = 2x$. Does that satisfy the definition of odd? Yes, it does, because $2(-x)$ is always equal to $-(2x)$. We can think about this graphically:

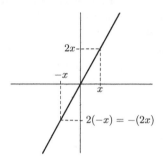

The same thing happens for the derivatives of other even functions (try it). As with the previous theorem, though, we could go on all day constructing even functions and differentiating them and checking that the derivative is odd, without getting any insight into *why* the conclusion always holds. For this theorem, you might be able to get an insight if you look at graphs of even functions and think about derivatives as gradients. Unfortunately, though, pointing at a graph ("Look!") isn't usually considered a good way of proving things. As before, we'll come back to this in Chapter 5.

4.4 A theorem with less familiar objects

We will now consider the fourth theorem, which will be harder because the objects might be entirely unfamiliar. If you have not yet taken a course called Linear Algebra, try to get a sense of this section but don't worry if you don't grasp the detail. The main point is that you should understand what objects the theorem is about—a map, sets of vectors, and three numbers—and how we might construct general representations of these.

(If it turns out that you don't understand this section at all, do continue to the rest of the chapter, as it deals with simpler ideas.)

Here is the theorem statement again:

Theorem: If $\phi : U \to V$ is a linear transformation, then
$$\dim(\ker(\phi)) + \dim(\mathrm{im}(\phi)) = \dim(U).$$

As usual, we will begin with the premise. Here the premise introduces an object $\phi : U \to V$, and tells us that this is a linear transformation. As before, the notation $\phi : U \to V$ means that ϕ is the name of a particular function that maps each element of the set U to an element of the set V. In this context U and V would be *vector spaces*, which are sets of vectors (with some particular properties). So we might imagine the situation using a diagram like this:

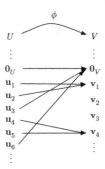

The curved arrow labeled "ϕ" indicates that ϕ is a map from U to V. The lower part of the diagram lists some actual vectors in U and V, and the straight arrows represent the way that ϕ sends particular vectors in U to particular vectors in V (for example, $\phi(\mathbf{u}_2) = \mathbf{v}_1$). I've used sets of three dots to indicate that there might be lots more vectors that are not shown. I've also included a zero vector for each of U and V (you will learn that every vector space has to have a zero vector). Notice that there's nothing to stop multiple vectors in U mapping to the same vector in V, and that it's not necessary that every vector in V gets mapped to.

Thus the diagram represents the situation described in the theorem premise. The conclusion says that

$$\dim(\ker(\phi)) + \dim(\operatorname{im}(\phi)) = \dim(U).$$

As in Chapter 2, "dim" is an abbreviation for *dimension* and "ker" is an abbreviation for *kernel*. Also, "im" is an abbreviation for *image*, and brackets used in expressions like these can be read as "of" (as is the case when we read $f(x)$ as "f of x"). So the conclusion can be read out loud as:

The dimension of the kernel of phi plus the dimension of the image of phi is equal to the dimension of U.

We can relate this to our diagram by looking up some definitions. The definition of *kernel* says:

Definition: Suppose $\phi : U \rightarrow V$ is a linear transformation. Then $\ker(\phi) = \{\mathbf{u} \in U | \phi(\mathbf{u}) = \mathbf{0}_V\}$.

This means that $\ker(\phi)$ is the set of all vectors $\mathbf{u}$ in U such that $\phi(\mathbf{u}) = \mathbf{0}_V$; that is, all the vectors in U that ϕ maps to the zero vector in V. Our diagram shows some vectors in U that map to $\mathbf{0}_V$, so these would all be in $\ker(\phi)$.

The definition of image says:

Definition: Suppose $\phi : U \rightarrow V$ is a linear transformation. Then $\operatorname{im}(\phi) = \{\mathbf{v} \in V | \mathbf{v} = \phi(\mathbf{u}) \text{ for some } \mathbf{u} \in U\}$.

This means that $\operatorname{im}(\phi)$ is the set of all vectors $\mathbf{v}$ in V such that $\mathbf{v}$ is equal to $\phi(\mathbf{u})$ for some vector $\mathbf{u}$ in U; that is, all the vectors in V that get "hit" by something coming from U. Our diagram shows some vectors in V that get hit by something, so these would be in $\operatorname{im}(\phi)$.

So $\ker(\phi)$ is a set of vectors in U, and $\operatorname{im}(\phi)$ is a set of vectors in V. Recall that U is a set of vectors too, and the conclusion relates the *dimensions* of each of these sets of vectors. Setting up a definition of *dimension* requires more work, but (as in Section 2.4), dimension can be described informally as the number of vectors you need to be able to "build" all the others. This means that $\dim(\ker(\phi))$, $\dim(\operatorname{im}(\phi))$, and $\dim(U)$ are all numbers, and the conclusion of this theorem is a statement relating these numbers.

This explanation involved a lot of objects and you might have found it rather abstract. When you study this topic in full, you will learn about lots of specific examples of vector spaces and linear transformations, so you will be able to think about it in terms of those examples as well as in terms of these general representations.

4.5 Logical language: "if"

I find that thinking about objects often allows me to link new statements to my existing knowledge. However, we can also work towards understanding statements by thinking about their logical structures. To do so, we need to pay attention to mathematical uses of logical language, a process we will begin in the remainder of this chapter. We will start by looking more carefully at ways in which the word *if* is used in mathematics.

First, consider a statement "If A then B." This is sometimes written as "$A \Rightarrow B$," which is read out loud as "A implies B." The use of an arrow makes it clearer that we can think of this statement as having a direction, which is important because we might have a situation in which $A \Rightarrow B$ is a true statement but $B \Rightarrow A$ is not. Sometimes both implications do hold, as in this case:

x is even $\Rightarrow x^2$ is even. (true)
x^2 is even $\Rightarrow x$ is even. (true)

However, sometimes one is true but the other is not, as in this case:

$x < 2 \Rightarrow x < 5$. (true)
$x < 5 \Rightarrow x < 2$. (false)

These statements are actually somewhat imprecise, because we have not specified what type of object x is. You probably assumed that it must be an integer in the squaring case and a real number in the inequalities case, since that would make sense. But, to be clearer, we could write things like:

For every $x \in \mathbb{R}, x < 5 \Rightarrow x < 2$.

Doing so makes it easier to see why this particular statement is false: there are some real numbers that are less than 5 but not less than 2. But mathematicians sometimes omit such phrases when writing in note form or when the intended interpretation is obvious.

When they want to discuss both implications at once, mathematicians use a double-headed arrow meaning "is equivalent to," or write "if and only if" or its abbreviation "iff." So these are different ways of writing the same (true) statement about integers:

x is even $\Leftrightarrow x^2$ is even.
x is even if and only if x^2 is even.
x is even iff x^2 is even.

We have seen the phrase "if and only if" before, in our definitions. Here is one again:

Definition: A number n is *even* if and only if there exists an integer k such that $n = 2k$.

To think about the phrase, it might be illuminating to split up this definition and write each implication separately:

A number n is even *if* there exists an integer k such that $n = 2k$.
A number n is even *only if* there exists an integer k such that $n = 2k$.

This is related to my point in Chapter 3 that we want our definition to "catch" the numbers that are even, and to exclude those that aren't. Can you see how?

One final thing to note about a statement of the form "A if and only if B" is that, if we want to prove it, we can take one of two approaches. We can either construct a proof in which all the lines are equivalent to each other, or prove the two statements $A \Rightarrow B$ and $B \Rightarrow A$ separately. Issues about proof construction and writing are discussed in more detail in Chapters 6 and 8.

4.6 Logical language: everyday uses of "if"

The uses of "if" and "$\Rightarrow$" sound straightforward when we are considering simple mathematical statements like those in the previous section. But I would like to draw your attention to two potential sources of confusion.

First, some thought is necessary to sort out which of the "if" and the "only if" corresponds to which implication. You should think about this, perhaps by thinking about which one could replace the "implies" arrow in these two versions of the same (true) statement:

$$x < 2 \Rightarrow x < 5 \qquad x < 5 \Leftarrow x < 2.$$

Second, it turns out that people do *not* always interpret "if" in a mathematical way in everyday life. In everyday conversation, we tend to speak rather imprecisely, relying on the context to help our listener make the interpretation we intend. For instance, imagine someone says to you,

If you clean the car then you can go out on Friday night.

You could reasonably infer from this that if you don't clean the car then you can't go out on Friday night. Clearly that is what the speaker intends. And someone else could infer that if you were allowed out on Friday, then you must have cleaned the car. But, in fact, *neither of these is logically equivalent to the original statement.* Perhaps the easiest way to see this is to look at the logic in parallel with our simple statements about inequalities:

clean car	$\Rightarrow$	out Friday	$x < 2$	$\Rightarrow$	$x < 5$
not clean car	$\Rightarrow$	not out Friday	$x \geq 2$	$\Rightarrow$	$x \geq 5$
clean car	$\Leftarrow$	out Friday	$x < 2$	$\Leftarrow$	$x < 5.$

The second and third statements in each case are not logically the same as the first. Alternatively, you could think about the everyday situation, and see that the original statement says nothing at all about what happens if you don't clean the car, so there would be no contradiction if you didn't clean it but you were still allowed out. Technically, the person bargaining with you should really say:

You can go out on Friday night *if and only if* you clean the car.

Of course, no-one talks like that. Which means that you might have less practice than you think at accurately interpreting logical statements. Using logical language in a mathematically correct sense is not too hard, though, because there are cases in which the intended natural language interpretation *is* the same as the mathematical one. Consider this statement:

If John is from Cardiff then John is from Wales.

No-one hearing this would dream of inferring either of these:

If John is not from Cardiff then John is not from Wales.
If John is from Wales then John is from Cardiff.

But these inferences are analogous to those we looked at for the statement about cleaning the car. Make sure you can see how.

In the Cardiff case, the natural interpretation of the statement is the same as the mathematical one. We can also use it to illustrate a general point about logical equivalence of different implications. For any statement of the form $A \Rightarrow B$ we can consider three related statements called its *converse*, *inverse*, and *contrapositive*. This is what each one means, using the Cardiff example for illustration:

original	$A \Rightarrow B$	from Cardiff $\Rightarrow$ from Wales
converse	$B \Rightarrow A$	from Wales $\Rightarrow$ from Cardiff
inverse	not $A \Rightarrow$ not B	not from Cardiff $\Rightarrow$ not from Wales
contrapositive	not $B \Rightarrow$ not A	not from Wales $\Rightarrow$ not from Cardiff.

This is what each one means using one of our simple mathematical examples instead:

original	$A \Rightarrow B$	$x < 2 \Rightarrow x < 5$
converse	$B \Rightarrow A$	$x < 5 \Rightarrow x < 2$
inverse	not $A \Rightarrow$ not B	$x \geq 2 \Rightarrow x \geq 5$
contrapositive	not $B \Rightarrow$ not A	$x \geq 5 \Rightarrow x \geq 2.$

This should help you remember that if $A \Rightarrow B$ is a true statement, then its contrapositive will also be true (in fact, they are logically equivalent), but its inverse and its converse might not be.

Finally, I should point out that your professor will be careful about uses of "if" and "if and only if" in theorems and proofs, but perhaps not so careful in definitions. In a definition, they might just write "if" instead of "if and only if." This works for the same reason that everyday communication works: everyone knows that, in a definition, this is what is intended.

4.7 Logical language: quantifiers

Next I want discuss the phrases "for all" and "there exists." These are called *quantifiers*, because they tell us how many of something we're talking about. They are so common in mathematics that we have symbols for them: we use "∀" (the universal quantifier) to mean "for all" and "∃" (the existential quantifier) to mean "there exists."

In simple statements, quantifiers are easy to think about. Here is a simple quantified statement:

$$\forall x \in \mathbb{Z}, \ x^2 \geq 0.$$

In this statement, we write $\forall x \in \mathbb{Z}$ to specify exactly which objects we are talking about. We could just write $\forall x$, and sometimes people do when it is obvious what kind of numbers (or other objects) a statement is about. But doing so could be ambiguous. In this case, the statement could just as well be about real or complex numbers, which raises issues of the truth or otherwise of the statement: "$\forall x \in \mathbb{Z}, \ x^2 \geq 0$" is true, "$\forall x \in \mathbb{C}, \ x^2 \geq 0$" is not. So it is good practice to be specific.

More complicated quantified statements can be harder to think about, although we had some practice with the definitions in Chapter 3. Here is one of these, written in words and in an abbreviated[3] form using the new symbol (which might be more naturally read as "for every" in this case):

Definition: A function $f : \mathbb{R} \to \mathbb{R}$ is *increasing* if and only if for every $x_1, x_2 \in \mathbb{R}$ such that $x_1 < x_2$, we have $f(x_1) \leq f(x_2)$.

Definition (abbreviated): $f : \mathbb{R} \to \mathbb{R}$ is *increasing* if and only if
$$\forall x_1, x_2 \in \mathbb{R} \text{ such that } x_1 < x_2, f(x_1) \leq f(x_2).$$

Here is a simple quantified statement involving the existential quantifier:

$$\exists x \in \mathbb{Z} \text{ such that } x^2 = 25.$$

This is a true statement, because when mathematicians say "there exists," they mean "there exists at least one." Here, there are two different integers x that satisfy the statement. In other cases, there might be hundreds. Students sometimes find it strange that we say "there exists" without

---

[3] If you're wondering why I haven't abbreviated further using ⇔ instead of "if and only if," it's because people don't tend to do that in definitions. I don't know why.

specifying how many because, if we know exactly how many there are, it seems rude not to say so. However, advanced mathematics is at least partly about general relationships between concepts, rather than about finding "answers" as such. We can see other reasons why it makes sense to use "there exists" without extra specification if we look at another of our definitions (again shown both in words and in an abbreviated form):

Definition: A number n is *even* if and only if there exists an integer k such that $n = 2k$.

Definition (abbreviated): $n \in \mathbb{Z}$ is *even* if and only if $\exists k \in \mathbb{Z}$ such that $n = 2k$.

This definition gives us an agreed way of deciding whether a number is even or not. To make that decision, we do not care what the particular k is, we just care whether or not there is one. It also allows us to make general arguments about all even numbers. We might start by saying "Suppose n is even, so $\exists k \in \mathbb{Z}$ such that $n = 2k$" (compare with our argument in Section 3.5). In this case, we don't want to specify what the k is because we want the ensuing argument to be general in the sense that it applies to any number that satisfies the condition.

4.8 Logical language: multiple quantifiers

Some mathematical statements have more than one quantifier. The following definition (again with an abbreviated version) might be described as doubly quantified or as having two nested quantifiers:

Definition: A set $X \subseteq \mathbb{R}$ is *open* if and only if for every $x \in X$ there exists $d > 0$ such that $(x - d, x + d) \subseteq X$.

Definition (abbreviated): $X \subseteq \mathbb{R}$ is *open* if and only if $\forall x \in X \, \exists d > 0$ such that $(x - d, x + d) \subseteq X$.

When a statement has more than one quantifier, *the order in which they appear is really important*. In this case, the definition says "for every x, there exists a d" and in Chapter 3, we imagined taking a particular x value and finding an appropriate d. We know that if we had taken a different x, we might have needed a different d (perhaps a smaller one, as in the right-hand diagram below).

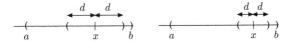

If the definition instead said "there exists a d for every x," mathematicians would read that as meaning that we could select a single d that works for every x. That is not the same thing at all.

To check your understanding of this, consider the following two statements. One is true, and the other is false. Which is which?

$$\exists y > 0 \text{ such that } \forall x > 0, y < x.$$

$$\forall x > 0 \; \exists y > 0 \text{ such that } y < x.$$

It is normal to find this difficult, because in everyday life we would probably not distinguish between these two statements. Without even realizing it, we would make the interpretation that seems most realistic, disregarding logical correctness. Most people therefore have to concentrate for a while before they get the hang of reading what is literally there and making the mathematically correct interpretation. (If you are unsure which statement is true, think about it some more then ask a tutor or professor.)

Using symbols like "$\Rightarrow$", "$\forall$", and "$\exists$" often shortens our mathematical sentences. I find it easier to work with something that looks shorter on the page, so I have always liked these symbols, and I tend to introduce them early on in my teaching. Other professors think that learning the new symbols interferes with students' concentration on the mathematical ideas, so they tend to write everything out in words. Still others write most things in words but make other types of abbreviation. Sometimes this is just a style issue, but sometimes it is a mathematical one—different symbols or abbreviations highlight different aspects of the mathematics, and some aspects might be considered worthy of detailed study in some courses but not in others. So you will probably see a variety of writing styles during your major. My advice would be to pay attention to what style the course professor seems to favor, and to adopt that style without being obsessive about it. At the end of the day, mathematical correctness and clarity is what counts, so you can abbreviate differently provided that your arguments are sound and well-written.

4.9 Theorem rephrasing

Although I have written the theorems in this chapter in the form "If...then...," you might also see different phrasings. Here are some common ones:

Theorem: If f is an even function, then $\dfrac{df}{dx}$ is an odd function.

Theorem: Suppose that f is an even function. Then $\dfrac{df}{dx}$ is an odd function.

Theorem: Every even function has an odd derivative.

These would all be interpreted to mean the same thing: the premise in each case is that the function is even, and the conclusion is that its derivative is odd. It might seem strange that we don't just pick one form and stick to it but, sometimes, one version or another sounds more natural, so mathematicians like to have this flexibility.

You will also see theorems of different types. There are, for instance, *existence theorems* like this one:

Theorem: There exists a number x such that $x^3 = x$.

One way to prove a theorem like this is just to produce an object that satisfies it: the number 1 would do, in this case. It's not always that easy, but it's important to recognize that it might be, because sometimes students tie themselves in knots doing complicated things when a simple answer would do.

There are also theorems about non-existence, like this one:

Theorem: There does not exist a largest prime number.

In fact, with a bit of thought, non-existence theorems can be restated in our standard form. This one, for instance, could be written with a universal quantifier:

Theorem: For every prime number n there exists a prime number p such that $p > n$.

Then it could be rephrased into our initial form:

Theorem: If n is a prime number then there exists another prime number p such that $p > n$.

These rephrasing possibilities can be very useful when we want to start proving something: sometimes rewriting in a different way can give us different ideas about sensible things to try. However, the fact that we can often rephrase does *not* mean that you can be sloppy about your mathematical writing. Sloppy paraphrasing can easily change the logical meaning of a statement. Once you become fluent in using logical language in a mathematical way, you will find that you can switch forms without doing violence to the meaning. Until that point you should think carefully about logical precision.

4.10 Understanding: logical form and meaning

One great thing about having precise meanings for logical terms is that it buys us a lot of mechanistic reasoning power. For instance, if we know that $A \Rightarrow B$ and that $B \Rightarrow C$, then we can deduce that $A \Rightarrow C$. We can do this even if A, B, and C are about really complicated objects that we've never met before and we don't understand. Similarly, if we would like to prove a statement of the form $A \Rightarrow B$, but we are not making much progress, we can remember that the contrapositive (not $B \Rightarrow$ not A) is always equivalent to the original, and try proving that instead.

This is what I mean when I say that we can develop valuable understanding by looking at the logical structure of a statement. If we pay attention to constructions involving "if" or quantifiers, we can make use of such regularities in our reasoning. This is (at least partly) what people mean when they talk about *formal* work: we can concentrate on the logical form of a sentence and temporarily ignore its meaningful content. We don't *have to* ignore the meaning, of course, and for most of the above discussion you were probably thinking about meanings as well. But attending to logical form is vital for proper understanding. Some students are lax about this; when reading mathematics, they look mostly at the symbols, ignoring or glossing over the words. This can make their understanding faulty and their writing inaccurate, because they mix up important quantifiers or implications. For instance, a question on one of my recent examinations required students to state this theorem:

Rolle's Theorem: Suppose that $f : [a, b] \to \mathbb{R}$ is continuous on $[a, b]$ and differentiable on (a, b) and that $f(a) = f(b)$. Then $\exists c \in (a, b)$ such that $f'(c) = 0$.

A few students made errors, writing things like this:

Rolle's Theorem: Suppose that $f : [a, b] \to \mathbb{R}$ is continuous on $[a, b]$ and differentiable on (a, b). Then $f(a) = f(b)$ and $\exists c \in (a, b)$ such that $f'(c) = 0$.

We can see that these are different by looking at their logical forms: one of the premises in the correct version appears instead as part of the conclusion in the second (look carefully to make sure that you can see this). Clearly that must make a very important difference, so the incorrect version cannot possibly be logically equivalent to the correct version. Nonetheless, it might still be a valid theorem. In this case, however, it isn't, which we can see by thinking in terms of examples. In the incorrect version, the premises introduce a function f that is defined on an interval and is continuous and differentiable on this interval. The conclusion claims that the function values are equal at the endpoints of the interval, but this cannot possibly follow in a valid way from the premises, because there are many functions and intervals that satisfy the premises but do not have this property. For example, $f(x) = x^2$ is continuous on $[0, 2]$ and differentiable on $(0, 2)$, but it certainly isn't the case that $f(0) = f(2)$. I can see how a student might write the incorrect version in the first place, but someone who is thinking about the meaning of their writing should recognize such errors when rereading.

This brings us back to the idea that both logical form and example objects can contribute to mathematical understanding, though focusing on each has different advantages and disadvantages. If you look mainly at examples, you might feel that you understand, but you might fail to appreciate the full generality of a statement or find it difficult to see the logical structure of a whole course. If you look mainly at formal arguments, you might be able to see how everything fits together logically, but you might find yourself complaining that it is very abstract and that you don't really understand what is going on. Your experience of these issues will probably vary from course to course, because some professors give lots of examples and draw lots of diagrams, whereas others give a much more formal presentation. If a professor's approach does not match your preferred way of developing understanding, you might find it useful to strengthen your understanding of the links between the example objects and the formal work. In this book, we have done a lot of that already; in the remainder of Part 1, we will shift our focus progressively towards working effectively with formal mathematics.

SUMMARY

- The premise of a theorem introduces the objects the theorem is about and tells us about their properties. The conclusion tells us something that follows logically from the premises (it doesn't tell us why it follows, just that it does).
- We can develop understanding of theorems by thinking about how they apply to examples. It is often a good idea to think about a range of examples; different representations might be useful too.
- Looking at lots of examples does not necessarily give any insight into why the conclusion of a theorem must always hold, which is what mathematicians are really interested in.
- Statements of the form "If A then B" are sometimes written as "$A \Rightarrow B$" ("A implies B"). Sometimes $A \Rightarrow B$ is true but $B \Rightarrow A$ is not. To talk about both directions, we say "if and only if" or use the symbol "$\Leftrightarrow$."

- In everyday life, people use "if" imprecisely; in mathematics they do not. It is important to be aware of this distinction if you want to interpret mathematical statements correctly.
- We use the symbol "∀" for "for all" and the symbol "∃" for "there exists." These are called quantifiers, and they appear frequently in mathematical statements. When more than one quantifier is used, their order is very important.
- Not all theorems are written in the form "If A then B." Some have different structures or are phrased differently. You should be careful when paraphrasing because a small change can dramatically alter the meaning.
- Understanding can be developed by looking at logical form and by examining how a theorem relates to example objects. These strategies are complementary and professors vary in which they emphasize.

FURTHER READING

For more on understanding the logical structures of mathematical statements, try:

- Houston, K. (2009). *How to Think Like a Mathematician*. Cambridge: Cambridge University Press.
- Allenby, R.B.J.T. (1997). *Numbers & Proofs*. Oxford: Butterworth Heinemann.
- Vivaldi, F. (2011). *Mathematical Writing: An Undergraduate Course*. Online at http://www.maths.qml.ac.uk/~fv/books/mw/mwbook.pdf.
- Velleman, D.J. (2004). *How to Prove It: A Structured Approach*. Cambridge: Cambridge University Press.
- Epp, S.S. (2004). *Discrete Mathematics with Applications*. Belmont, CA: Thompson-Brooks/Cole.

CHAPTER 5

Proof

This chapter is about why mathematicians make the effort to prove things, and how they do so. It describes common undergraduate proof tasks and useful strategies for tackling these. It also explains why mathematicians sometimes ask students to prove obvious things, and how to respond when you come across a result that seems obviously true but that turns out to be false.

5.1 Proofs in high school mathematics

You have been constructing mathematical proofs for many years. For instance, to do the calculations needed to prove that the solutions to the equation $x^2 - 20x + 10 = 0$ are $10 + 3\sqrt{10}$ and $10 - 3\sqrt{10}$, you would write something like this:

$$x = \frac{20 \pm \sqrt{400 - 40}}{2} = \frac{20 \pm \sqrt{360}}{2} = \frac{20 \pm 6\sqrt{10}}{2} = 10 \pm 3\sqrt{10}.$$

This calculation uses methods that everyone agrees are valid, so it captures everything we need for a proof that the solutions really are as claimed. To make it look more like an upper-level proof, we could rewrite it like this:

Claim: If $x^2 - 20x + 10 = 0$ then $x = 10 + 3\sqrt{10}$ or $x = 10 - 3\sqrt{10}$.

Proof: Suppose that $x^2 - 20x + 10 = 0$.

Then, using the quadratic formula,

$$x = \frac{20 \pm \sqrt{400 - 40}}{2}$$

$$= \frac{20 \pm \sqrt{360}}{2}$$

$$= \frac{20 \pm 6\sqrt{10}}{2} \quad \text{(because } \sqrt{ab} = \sqrt{a}\sqrt{b}\text{)}$$

$$= 10 \pm 3\sqrt{10}.$$

This version explicitly states the claim, begins with the premise ("Suppose that $x^2 - 20x + 2 = 0$"), and contains a few words to justify those steps that are more sophisticated or less obvious.

My point is that there is nothing inherently mysterious about proofs. It is certainly true that most high school mathematics is not presented in this way, but it's also true that most of it could be. I say this because, in upper-level courses, lots of the mathematics you meet will be presented in this form; your lecture notes will be full of theorems and proofs. This might seem rather an abrupt change, and some students get the idea that proof is a mysterious dark art to which only the very privileged have access. It isn't. In cases like the one above, all it really involves is writing in a more mathematically professional way—writing less like a student and more like a textbook, if you like.

This is not to belittle the genuine difficulties that undergraduate students face when handling proofs. Obviously the example above is a simple one, and the proofs you are asked to understand and construct in upper-level courses will often (though not always) be much harder. It might take you a while to get used to digesting mathematics presented in this way, and to writing your own mathematics more professionally. But there is no reason to think you won't manage it, and this chapter discusses some things you could pay attention to in order to get used to it quickly.

5.2 Proving that a definition is satisfied

One thing you'll often be asked to do is to prove that a mathematical object satisfies a definition. The question won't be phrased like that, though. It will just say something like, "Prove that the set $(2, 5)$ is open." You will have to interpret this to mean "Prove that the set $(2, 5)$ satisfies the definition of open," and review the definition of open set to make sure you are clear about what this involves. That sounds simple, but I have often seen students who, faced with an instruction like "Prove that the set $(2, 5)$ is open," don't know what to do. If you don't know how to start on any proof problem, your first thought should often be, what does the definition say?

We have seen the relevant definition, which says:

Definition: A set $X \subseteq \mathbb{R}$ is *open* if and only if $\forall x \in X \, \exists d > 0$ such that $(x - d, x + d) \subseteq X$.

In Chapter 3 we convinced ourselves that $(2, 5)$ does satisfy this definition. But how should we write a proof? Often the best advice is to follow the structure of the definition itself. We want to prove that $(2, 5)$ is open, so we want to show that for every $x \in (2, 5)$ there exists $d > 0$ such that $(x - d, x + d) \subseteq (2, 5)$. When we want to show something is true for every x in some set, we usually start our proof by introducing one, like this:

Claim: $(2, 5)$ is an open set.

Proof: Let $x \in (2, 5)$ be arbitrary.

Here, *arbitrary* means that we are taking any old x, not some specific one or one with special properties. It is not necessary to write this—lots of people would just write "Let $x \in (2, 5)$"—but it does emphasize that the ensuing argument will work for any x in the set.

Now we need to show the existence of an appropriate d. In Chapter 3, we convinced ourselves that this was possible by thinking about a diagram. That's fine for informal understanding, but when writing proofs we should be more precise. Doing so need not be complicated, because the simplest way to show the existence of something is to produce one. In this case, we need to produce a d that will work for our x. This d will depend on x, and one way to do it is to pick d to be the minimum of the two distances $5 - x$ and $x - 2$ (think about why). So we might write the rest of our proof like this:

Claim: $(2, 5)$ is an open set.

Proof: Let $x \in (2, 5)$ be arbitrary.

Let d be the minimum of $5 - x$ and $x - 2$.

Then $(x - d, x + d) \subseteq (2, 5)$.

So $\forall x \in (2,5) \; \exists d > 0$ such that $(x - d, x + d) \subseteq (2,5)$

So $(2,5)$ is an open set.

Notice that this proof reflects the order and structure of the definition. We are showing something for all x, so we start with an arbitrary one. We are showing that for this x, an appropriate d exists, so we produce one. Because of this, the structure of the proof will be obvious to a mathematician, so you don't really need to write the line "So $\forall x \in (2,5)\ldots$," but you might find it helpful for your own thinking.

Some people tend to include diagrams along with proofs like this, and some don't. That's because diagrams can be illuminating, but they are not necessary, and they are not a substitute for writing a proof out properly; mathematicians want to see a written argument that is clearly linked to the appropriate definition. That said, I have been known to give points in an exam when someone has shown some understanding by drawing a diagram, even if they haven't succeeded in turning this into a full written proof. The extent to which this happens will probably depend on the subject matter and on the way your professor wants to balance evidence of conceptual understanding with evidence of ability to write good proofs.

5.3 Proving general statements

Mathematics professors will want you to construct proofs that specific objects satisfy specific definitions, and they will want you to find solutions to standard problems or equations, as in the quadratic formula illustration. But they are often more interested in general results: in theorems that hold for a whole class of objects. Fortunately, you've had experience of the reasoning needed for this type of proof, too. For instance, you might have seen an exercise like this:

Show that $\cos(3\theta) = 4\cos^3\theta - 3\cos\theta$.

The "show" is a big giveaway—mathematicians use it as a synonym for "prove." And notice that, although this is not explicitly stated, the idea is to show that the equation holds for every possible value of θ. The exercise can be tackled using the standard trigonometric identities, and in answering it you would probably write a series of equations. Again, these could be written in the form of a theorem and proof:

Theorem: For every $\theta \in \mathbb{R}$, $\cos(3\theta) = 4\cos^3\theta - 3\cos\theta$.

Proof: Let $\theta \in \mathbb{R}$ be arbitrary. Then

$$\begin{aligned}
\cos(3\theta) &= \cos(2\theta + \theta) \\
&= \cos(2\theta)\cos\theta - \sin(2\theta)\sin\theta \\
&= (\cos^2\theta - \sin^2\theta)\cos\theta - 2\sin\theta\cos\theta\sin\theta \\
&= \cos^3\theta - 3\sin^2\theta\cos\theta \\
&= \cos^3\theta - 3(1 - \cos^2\theta)\cos\theta \\
&= 4\cos^3\theta - 3\cos\theta, \text{ as required.}
\end{aligned}$$

Because this argument holds for every value of θ, we have proved an *identity* (we say something is *identically equal* to something else if it is equal in all cases). When mathematicians want to draw attention to the fact that an equation is an identity, they sometimes use an equals sign with an extra bar, like this: $\cos(3\theta) \equiv 4\cos^3\theta - 3\cos\theta$. All of the equals signs in the proof above could be replaced with this symbol, but instead the proof makes the generality clear by starting with "Let $\theta \in \mathbb{R}$ be arbitrary." Incidentally, you don't need to write "as required" at the end. Sometimes people just stop, sometimes they draw a little square, and sometimes they write "QED."

Because it applies to all values of θ, this argument is logically different from what we did with the quadratic equation—we can write down "$x^2 - 20x + 10 = 0$" but we know that there are lots of values of x for which this does not hold. The generality of the theorem we just proved makes it more like those in Chapter 4, and more mathematically interesting. We can, however, look at a similar level of generality by examining the quadratic formula itself. What's really mathematically interesting about the quadratic formula is that it *always* works. When you stop to think about it, this is very powerful. We have a simple formula that allows us to find solutions to infinitely many equations. When we want to solve one, we just substitute the appropriate numbers into the formula, and 30 seconds later we know the answers. I realize that you are completely accustomed to mathematics being like this—you know a whole host of formulas for all kinds of things, so you are probably not amazed by it at all. But someone had to discover those formulas. The first people to do this didn't read them in a textbook or get told them by a teacher—they worked them out for themselves. In the case of the quadratic formula, explicit solutions for some types of quadratic equation appeared in the seventh century, and the formula as we know it appeared in the seventeenth century. So, when you solve a quadratic equation you are using knowledge that has only been available to humanity for around 400 years. And much of what you'll learn as an undergraduate is much newer.

In any case, in advanced mathematics there is less emphasis on using formulas to do specific calculations, and more emphasis on understanding where those formulas come from and proving that they always work. You might already have seen a proof that the quadratic formula works. But, if you tend to ignore things that will not be on an exam, you might not have noticed, so we will look at one now. We stated the appropriate theorem in Chapter 4:

Theorem: If $ax^2 + bx + c = 0$, then $x = \dfrac{-b \pm \sqrt{b^2 - 4ac}}{2a}$.

In fact, we can improve this by specifying that x is a complex number, and write:

Theorem: Suppose that $x \in \mathbb{C}$.

If $ax^2 + bx + c = 0$, then $x = \dfrac{-b \pm \sqrt{b^2 - 4ac}}{2a}$.

Proof: Suppose that $x \in \mathbb{C}$. Then

$$ax^2 + bx + c = 0 \Rightarrow \left(\sqrt{a}x + \frac{b}{2\sqrt{a}}\right)^2 - \frac{b^2}{4a} + c = 0$$

(completing the square)

$$\Rightarrow \left(\sqrt{a}x + \frac{b}{2\sqrt{a}}\right)^2 = \frac{b^2}{4a} - c$$

$$\Rightarrow \sqrt{a}x + \frac{b}{2\sqrt{a}} = \pm\sqrt{\frac{b^2}{4a} - c}$$

$$\Rightarrow \sqrt{a}x + \frac{b}{2\sqrt{a}} = \pm\sqrt{\frac{b^2 - 4ac}{4a}}$$

$$\Rightarrow \sqrt{a}x = -\frac{b}{2\sqrt{a}} \pm \sqrt{\frac{b^2 - 4ac}{4a}}$$

$$\Rightarrow x = -\frac{b}{2a} \pm \frac{1}{\sqrt{a}}\sqrt{\frac{b^2 - 4ac}{4a}}$$

$$\Rightarrow x = -\frac{b}{2a} \pm \frac{1}{\sqrt{a}}\frac{1}{2\sqrt{a}}\sqrt{b^2 - 4ac}$$

$$\Rightarrow x = \frac{-b \pm \sqrt{b^2 - 4ac}}{2a}, \text{ as required.}$$

One thing to notice about this proof is that it differs from that for the trigonometric identity in the way we use the symbols "$=$" and "$\Rightarrow$", and that this reflects the structure of what we are proving. For the trigonometric identity, we were proving that an equation always holds. Thus we began with one side of the equation and showed that we could get to the other side through a series of equalities. For the quadratic formula, we are proving that *if* the original equation holds, *then* x must take on certain values. Thus we start with the whole equation, and show at each step that the truth of the current equation implies the truth of the next one.

Another thing to notice is that you probably found yourself reading back and forth over parts of the proof a few times. In particular, you might have found it easier to see that

$$ax^2 + bx + c = 0 \Rightarrow \left(\sqrt{a}x + \frac{b}{2\sqrt{a}}\right)^2 - \frac{b^2}{4a} + c = 0$$

by going from right to left (multiplying out) rather than from left to right. Indeed, it may be easier to check the algebraic manipulations by reading from bottom to top than from top to bottom. That might make you wonder why we write it this way around. The reason is that we want the logic to "flow" in relation to the whole theorem. The theorem says that if $ax^2 + bx + c = 0$, then the conclusion about the x values follows. We want our proof to reflect that structure: to start with the premise and to proceed by taking valid steps until we reach the conclusion. We will come back to issues about order in proofs in Chapter 8.

5.4 Proving general theorems using definitions

Another common mathematical task is to prove that if one definition is satisfied, then so is another one. Quite a few theorems amount to a statement that can be proved like this, so it is often possible to make progress by writing down what the premises mean in terms of definitions, writing down what the conclusion means in terms of definitions, then working out how to get between the two.

Here we will look at an illustration that uses the definition of the derivative, which was introduced in Chapter 4:

Definition: $\dfrac{df}{dx} = \lim\limits_{h \to 0} \dfrac{f(x+h) - f(x)}{h}$, provided this limit exists.

Using this notation, the *sum rule* for derivatives could be stated like this:

Theorem: Suppose that f and g are both differentiable. Then

$$\frac{d}{dx}(f+g) = \frac{df}{dx} + \frac{dg}{dx}.$$

I realize that you probably think of this theorem as obviously true. If so, you should ask yourself whether you have a good reason for thinking that, or whether you've just got used to it. Proving it amounts, as always, to showing that the conclusion follows from the premise. In this case, the premise is that f and g are both differentiable, which means that they both have derivatives, which means that their derivatives are defined by the definition. That is, we will assume that

$$\frac{df}{dx} = \lim_{h \to 0} \frac{f(x+h) - f(x)}{h} \quad \text{and that} \quad \frac{dg}{dx} = \lim_{h \to 0} \frac{g(x+h) - g(x)}{h}.$$

We'd like to prove that the conclusion holds, that is, that

$$\frac{d}{dx}(f+g) = \frac{df}{dx} + \frac{dg}{dx}.$$

Notice that on the left, $f + g$ is treated as a single function that will be differentiated. On the right, each of the functions f and g is differentiated separately, and the resulting derivatives are added together. The notation thus reflects a difference in the order of the addition and differentiation operations, and the theorem really says that this order does not matter: we get the same thing both ways.

To get started on a proof, we can state the conclusion more precisely in terms of the definitions too:

$$\lim_{h \to 0} \frac{(f + g)(x + h) - (f + g)(x)}{h}$$
$$= \lim_{h \to 0} \frac{f(x + h) - f(x)}{h} + \lim_{h \to 0} \frac{g(x + h) - g(x)}{h}.$$

Hopefully you can see that there isn't much to do here, because it is a very short step from the premises to the conclusion. We can write something like this:

Theorem: Suppose that f and g are both differentiable. Then

$$\frac{\mathrm{d}}{\mathrm{d}x}(f + g) = \frac{\mathrm{d}f}{\mathrm{d}x} + \frac{\mathrm{d}g}{\mathrm{d}x}.$$

Proof: Suppose that f and g are differentiable.

Then

$$\frac{\mathrm{d}}{\mathrm{d}x}(f + g) = \lim_{h \to 0} \frac{(f + g)(x + h) - (f + g)(x)}{h}$$
$$= \lim_{h \to 0} \frac{f(x + h) + g(x + h) - f(x) - g(x)}{h}$$
$$= \lim_{h \to 0} \left(\frac{f(x + h) - f(x)}{h} + \frac{g(x + h) - g(x)}{h} \right)$$
$$= \lim_{h \to 0} \frac{f(x + h) - f(x)}{h} + \lim_{h \to 0} \frac{g(x + h) - g(x)}{h}$$
$$= \frac{\mathrm{d}f}{\mathrm{d}x} + \frac{\mathrm{d}g}{\mathrm{d}x}, \text{ as required.}$$

Notice that the first and last steps make direct use of the equality in the definition; it is the intermediate steps that require some thinking. Notice also that when writing detailed proofs like this, we often perform only one manipulation at each step. There is a bit of leeway for obvious manipulations, but generally we are careful to avoid taking anything for granted. Your professor might present this kind of thing in a slightly different way, using alternative notation or considering the derivatives at a specific point rather than considering whole functions. The main point stands, though: we can often make progress by writing down the premises and conclusion in terms of the relevant definitions, then looking for a way to get between the two.

5.5 Definitions and other representations

Often, once everything is written down in terms of definitions, it is fairly clear how to get started on a proof. In some cases, though, a bit more creativity is required—you have to have a good idea. We've already looked at one way to get ideas: look at some example objects, perhaps using a variety of representations. Here I'll discuss this strategy in relation to some theorems from earlier chapters.

First, consider this theorem:

Theorem: If l, m, and n are consecutive integers, then the product lmn is divisible by 6.

In Chapter 4, we looked at objects that satisfy the premise (consecutive integers l, m, and n) and checked that they also satisfy the conclusion (that the product lmn is divisible by 6). But I commented that we could keep doing such multiplications all day without getting any insight into *why* the theorem must be true or how we might prove it.

To try to construct a proof, we can apply the strategy of stating everything in terms of definitions. Here the premise is that the three integers are consecutive, which is not obvious from the notation l, m, and n. We could call our numbers l, $l + 1$, and $l + 2$ instead, which gives us the option of multiplying them together and seeing what we get:

$$l(l + 1)(l + 2) = l(l^2 + 3l + 2) = l^3 + 3l^2 + 2l.$$

I don't know about you, but that doesn't help me much. I can't rearrange the last expression in any obvious way that will allow me to pull out a factor of 6. It was worth a try though—sometimes such manipulations will reveal an obvious factor.

We can, however, try working with the conclusion as well. In this case, we want to prove that lmn is divisible by 6 which, thinking in terms of prime factorizations, means that it has to be divisible by 2 and divisible by 3. This is more helpful, and for me the number line is a useful representation for thinking about it.

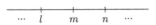

Looking at the number line, I can "see" that at least one of the numbers l, m, and n must be divisible by 2—every second number is even, so either l and n are both even or m is even. And exactly one of l, m, and n must be a multiple of 3, for a similar reason. So the number lmn must be a multiple of 3. This explains why the result is true, and we might write a proof like this:

Theorem: If l, m, and n are consecutive integers, then the product lmn is divisible by 6.

Proof: Suppose that l, m, and n are consecutive integers.

Then at least one of l, m, and n is divisible by 2.

Also, exactly one of l, m, and n is divisible by 3.

Hence the product lmn is divisible by 2 and is divisible by 3.

So the product lmn is divisible by 6.

This is a perfectly good proof. Sometimes students think it isn't—they think it isn't long enough, or isn't complicated enough, or doesn't contain enough algebra. But a proof is just a logical argument that makes clear why a theorem is true. Sometimes such arguments have to be complicated because a lot of logical links are involved. But sometimes there aren't many links, and a proof can be very simple.

Next, consider this theorem:

Theorem: If f is an even function, then $\dfrac{df}{dx}$ is an odd function.

In Chapter 4 I suggested that thinking about this theorem in terms of a graph might give you some insight into why it is true. Here is a diagram with some extra labels that might help:

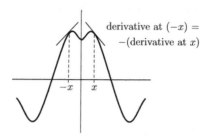

derivative at $(-x) =$
$-(\text{derivative at } x)$

You might find this pretty compelling evidence. But it isn't enough for a proof, because a proof is supposed to show how a theorem follows deductively from definitions or other accepted results. In those terms, the premise means that for every $x, f(-x) = f(x)$, and the conclusion means that $f'(-x) = -f'(x)$ (I find it easier to use this alternative derivative notation for this kind of question). We would like to go from the premise to the conclusion, and the obvious way to do this is to differentiate both sides of the equation $f(-x) = f(x)$. It's not quite clear in advance whether that will work, but we have to do something and it is the obvious thing to try.

The derivative of the right-hand side is clearly just $f'(x)$. It might be obvious to you that the derivative of the left-hand side is $-f'(-x)$ by the chain rule. If not, it might help to notice that $f(-x) = f(g(x))$ where $g(x) = -x$, so $f'(-x) = f'(g(x))g'(x) = f'(-x) \cdot (-1) = -f'(-x)$. In any case, a bit of rearranging does indeed give us the desired conclusion. Then there is the question of writing the argument clearly, which we might do like this:

Theorem: If f is an even function, then f' is an odd function.

    Proof: Suppose that f is an even function.

        Then for every x we have $f(-x) = f(x)$.

        Differentiating both sides gives $-f'(-x) = f'(x)$.

So, for every $x, f'(-x) = -f'(x)$.

So f' is an odd function, as required.

5.6 Proofs, logical deductions, and objects

In the example about even functions and their derivatives, you might have looked at the proof and thought, why bother thinking in terms of a diagram, then? The proof uses straightforward algebra and known rules—surely it's quicker just to use those straightaway? To this I would say: you're right—it is often very sensible to take that approach. In some cases using definitions and theorems and algebra and logic does lead very quickly to a proof. Moreover, it can be a distinct advantage *not* to think explicitly about other representations of the objects. Consider, for instance, the proofs we looked at near the beginning of this chapter about trigonometric identities and the quadratic formula. In those cases, you *could* have thought about examples—particular equations and solutions, or particular angles or cosine graphs—but you almost certainly didn't. You coped just fine with the abstract symbolic representations, so much so that you probably didn't feel that you were doing anything very abstract at all.

In fact, you can do even better than that: you can operate in situations where you have no idea what any of the objects are. If you're told that every bloober is a cobo and every cobo is a turit, you can deduce immediately that every bloober is a turit. You just used logical reasoning to do that. If you're told that bloobers are bigger than zocks and zocks are bigger than ngurns, you can confidently say that bloobers are bigger than ngurns. In that case, you used logical reasoning plus your knowledge about how "bigger than" works (in mathematical terms, the relation "is bigger than" is *transitive*). You have no idea what bloobers and cobos and turits and zocks and ngurns are, but this presents you with no problem whatsoever.

So it might well be that you can learn a lot of your undergraduate mathematics by paying attention solely to formal, logical arguments, without having to think explicitly about the objects referred to by the statements. Of course, you need to interpret and use logical language correctly: as discussed at the end of Chapter 4, you should take care with implications and quantifiers, and make sure that you don't inadvertently use the converse or the inverse of a theorem rather than the theorem itself. It is important to remember this because some students try to work formally but aren't very good at it. They don't have good logical reasoning skills (they make mistakes like those we'll discuss in more detail in Chapter 8) and they try manipulations at random, without much sense of whether these are likely to be productive and without much reflection on whether each step is valid. However, some people get really good at formal work, because they have good logical reasoning skills and because they develop a sort of "symbolic intuition" that tells them which manipulations are likely to be useful and which logical deductions are likely to lead to the required result. They might be especially good if they have paid attention to common proof structures within a particular course or topic (for more on this, see Chapter 6).

Personally, I *do* often like to think about the objects to which mathematical arguments apply. I find that diagrammatic representations in particular give me an intuitive sense that things must be the way they are, and I find this helpful for reconstructing definitions, theorems, or proofs that I can't quite remember. I'm aware, though, that it's not always easy to capture intuitive, object-based reasoning in formal terms. It takes some work to consider the full range of variability in the relevant objects and to translate intuitive understanding into the appropriate symbolic notation.

The upshot of this is that if you're expecting me to say that either formal or intuitive strategies are "best," you're going to be disappointed. Each has different advantages and disadvantages, and successful mathematicians are usually aware that both strategies exist. They might start with one—perhaps because of overall personal preference or because that's just how they tend to think for that particular mathematical topic—and try that for a bit, then switch to the other one if they get stuck. And if they get stuck with that one too, they might switch back, perhaps with some new insight that means the first strategy will now work better. We don't really know how they know when to switch (they probably don't know either—experts can be very bad at reflecting on the nature of their expertise). We do know that experts do not waste time bashing away at one particular strategy when it's clearly not getting them anywhere. They tend to recognize when they're in that situation, and do something to get out of it. That might mean switching strategy, or pausing to clarify the exact nature of the difficulty so that the existing strategy can be amended, or having a break and a cup of coffee, or deciding to talk it through with someone else. Being a good mathematical problem solver does involve being willing to persevere, but it also involves being able to recognize when an approach isn't working.

5.7 Proving obvious things

In the remainder of this chapter, we will look less at the mechanics of proving, and more at the issue of why we bother with it anyway (beyond the general satisfaction of putting together a logically correct argument).

Lots of introductory books on advanced mathematics will tell you that we prove things because we want to be absolutely certain that they are true. That is correct but it is a bit disingenuous, because mathematicians also ask students to prove things that they (the students) are already certain about. For instance, you are already certain that when we add together two even numbers, we get another even number. Seeing a proof of this will not make you any more certain of it, and rightly so. Sometimes mathematical claims that appear at first to be true do turn out not to be—we'll look at some of those in the next sections. So we do have to be careful. But you'll also find that as a mathematics major, professors will expect you to expend quite a bit of effort proving things that you already know. Indeed, mathematicians themselves collectively spend a fair amount of time in similar activity.

Some people will explain this by saying that professors want you to practice writing rigorous proofs of simple things, so that you can use your proving skills confidently when you go on to work with more complicated things. That is also true, and it's certainly a sensible reason to do it. It's still not the whole truth, however, because it doesn't explain why mathematicians do this kind of thing for themselves as well as making students do it.

A better reason is that mathematicians value not just knowing that a theorem is true, but also understanding how it fits into a broader network of connected results that forms a coherent theory. They don't just want to prove that the sum of two even numbers is even, they want to see how that is related to the way in which we prove that the sum of two odd numbers is even, or the way in which we prove that the sum of any two numbers that are divisible by 3 is also divisible by 3. They want to formulate the whole lot in similar terms, so that they can see relationships across a whole theory about, in this case, divisibility properties of numbers. Doing so entails specifying definitions from which we can build up lots of different proofs, and working out how to present those proofs so that the structural relationships between them are clear.

Indeed, there is more to the development of mathematics than just assembling a coherent theory. Ideally, mathematicians like to build up a very large theory on the basis of a very small number of initial axioms and definitions. This is seen as an important intellectual goal: we do not want to assume something as an axiom if, in fact, we can prove it as a theorem. This sounds like a clear and sensible aim, but some of its consequences can come across as confusing to new mathematics majors. To illustrate, at some point (probably in a course called something like Foundations or Analysis), a professor will present you with a list of axioms, definitions, and properties associated with the real numbers. This list will include things like this:

Definition: We say $a < b$ if and only if $b - a > 0$.

The professor will then ask you to prove things like this:

Claim: If $0 < -a$ then $a < 0$.

This might seem a bit weird. You will probably look at the axioms, definitions, and properties and think they're obvious, and then look at the thing you're asked to prove and think that's obvious too—just as obvious as the axioms, definitions, and properties, in fact. Such an experience has some entertaining emotional consequences: you'll see students going around being outraged and saying things like "I can't believe they're insulting my intelligence by asking me to prove things I've known since I was six!"

In fact, your professors are not asking you to prove this kind of thing to show that you know it. They know that you know it. They're asking you to prove it to show that you are disciplined enough in your thinking to treat a statement as a theorem that can be established on the basis of only a restricted set of axioms, definitions, and properties. It's actually quite difficult to exercise this discipline; when you try to construct such proofs, you will find yourself tempted to assume all sorts of other "obvious" things that aren't on the list. So try not to get impatient with tasks like this—if you think about them in the right way, they constitute an interesting and valuable intellectual exercise.

5.8 Believing counterintuitive things: the harmonic series

In the last section I mentioned that some mathematical results appear at first to be true but, on careful inspection, turn out not to be. These things are sometimes referred to as *counterintuitive*, meaning that they go against (most people's) intuition. Their existence is one reason why mathematicians take care to prove things properly: sometimes everyone believes something which later turns out to be just wrong. I will give a couple of my favorite illustrations in this section and the next, and will use them to make some general points about how you should respond to this kind of thing.

The first illustration involves the infinite sum

$$1 + \frac{1}{2} + \frac{1}{3} + \frac{1}{4} + \frac{1}{5} + \frac{1}{6} + \cdots .$$

This is called an *infinite series*, or sometimes just a *series*. One thing to note is that the ellipsis (the "...") is very important. It can be read as "and so on" and it means "and so on forever." If you omit it, mathematicians will think you mean to stop wherever you stop.

This particular series is called the *harmonic series*, and the obvious question is, what does it add up to? Mathematicians think about this by thinking about *partial sums*, so called because each one involves only a finite number of the terms. Here are the first few partial sums:

$s_1 = 1$
$s_2 = 1 + \frac{1}{2} = \frac{3}{2}$
$s_3 = 1 + \frac{1}{2} + \frac{1}{3} = \frac{11}{6}$
$s_4 = 1 + \frac{1}{2} + \frac{1}{3} + \frac{1}{4} = \frac{25}{12}.$

You might like to calculate a few more, to get a feel for what is happening. We could also represent these partial sums graphically, by plotting s_n against n, as in the graph below (notice, by the way, that this graph has dots and not a curve, because the sums are only defined for whole numbers):

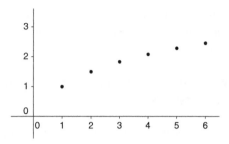

Now, suppose we go on forever. What do you think the infinite series will add up to?

Most people say something like "maybe 5" or "certainly less than 10." They are wrong. The answer is that the sum is infinitely large. It is bigger than 5, and bigger than 10, and bigger than any other number you care to name. We can demonstrate this as follows.

First, note that the first term, 1, is greater than $\frac{1}{2}$. The second term is equal to $\frac{1}{2}$. The third term is not greater than or equal to $\frac{1}{2}$ on its own, but if we take the next two terms we get something that is: $\frac{1}{3} + \frac{1}{4}$ is definitely bigger than $\frac{1}{2}$. So we are adding another half (plus a bit extra). Similarly, for the next four terms, $\frac{1}{5} + \frac{1}{6} + \frac{1}{7} + \frac{1}{8} > \frac{1}{2}$. So we are adding another half. And we can keep adding another half as many times as we like, by taking the next eight terms, then the next sixteen, then the next thirty-two, and so on. Sometimes people write things like this to represent the argument:

$$1 + \frac{1}{2} + \underbrace{\frac{1}{3} + \frac{1}{4}}_{>\frac{1}{2}} + \underbrace{\frac{1}{5} + \frac{1}{6} + \frac{1}{7} + \frac{1}{8}}_{>\frac{1}{2}} +$$

$$\underbrace{\frac{1}{9} + \frac{1}{10} + \frac{1}{11} + \frac{1}{12} + \frac{1}{13} + \frac{1}{14} + \frac{1}{15} + \frac{1}{16}}_{>\frac{1}{2}} + \cdots .$$

Because we can keep adding more halves, the sum is bigger than 5, and 10, and so on. We need an awful lot of terms before the total is bigger than, say, 100, but that's okay because we've got infinitely many of them.

In fact, that's one thing that you should learn from this: infinity is *really* big. The terms are really small, but there are so many of them that they add up to something infinitely big anyway. Most people find this surprising, and feel that they've learned something genuinely new by looking at the argument above. In fact, while you are a mathematics major, you will learn about infinite series that behave even more strangely than this one. They are called *conditionally convergent* series and (in my opinion) they're one of the most interesting things in undergraduate mathematics. Look out for them.

The other thing to learn is that things that are intuitively "obvious" sometimes turn out to be completely false, especially when they involve infinite processes. In this case, the intuition we got from looking at a few terms and drawing a graph was badly misleading. That doesn't mean you should stop using graphs or trusting your intuition; it just means you should do so with appropriate caution.

5.9 Believing counterintuitive things: Earth and rope

The next illustration uses different mathematics: simple geometry and algebra. I first saw this when I was a PhD student; a couple of my friends who were doing biology said that someone had told them this unbelievable thing, and asked me whether it was true. It turns out that it is, and I'll present it in the way I first saw it, as a question.

First, imagine the Earth, and imagine that instead of being all bumpy with mountains and so on, it is a nice, smooth sphere. Imagine that we get a rope and put it around the equator. We pull it so that it is taut; so that it fits snugly.

Now imagine that we add a meter—just one meter—to our rope. Then we shuffle it out all the way around the Earth, so that it is the same height above the surface at every point.

The question is: How far is the rope from the surface of the Earth?

Most people say "Oh, something really small, like a fraction of a millimeter." This is an intuitively natural response, because the Earth is really big, and a meter is really small by comparison, so it seems like adding a meter should make hardly any difference.

In fact, the answer is about 16 centimeters.

If you haven't seen this before, you will probably be skeptical about this answer. But it is correct, and the mathematics we need to prove this is very straightforward. Here it is, in the form of a proof with some accompanying diagrams.

Proof: We will work in meters throughout.

Let the radius of the Earth be denoted by R.

Then the circumference of the Earth is $C = 2\pi R$.

Adding a meter of rope gives a new circumference of $C' = 2\pi R' = 2\pi R + 1$.

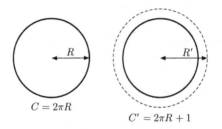

$C = 2\pi R$

$C' = 2\pi R + 1$

Solving to find the new radius R' we get

$$R' = \frac{C'}{2\pi} = \frac{2\pi R + 1}{2\pi} = \frac{2\pi R}{2\pi} + \frac{1}{2\pi} = R + \frac{1}{2\pi}.$$

Thus the new radius R' is equal to the original radius R plus $1/2\pi$ meters.

$1/2\pi \approx 0.16$, so the new radius is about 16 cm longer than the original, meaning that the rope is now 16 cm above the surface of the Earth.

Once again, the intuitive answer that most people give is completely wrong. This case seems worse though; it seems to shake people up because they were so certain about their first intuition. Some people even go so far as to insist that there must be a mistake in the proof. There isn't. It's completely correct, and you should (as always with proofs) check all the steps to make sure you are convinced of this.

So how should you respond to such a counterintuitive result? You could decide it's just too disturbing, and not give it any more thought. But that would be missing a valuable opportunity to develop your thinking. I'll tell you what I did, to show you what I mean. First, I convinced myself that the argument really did work—I checked carefully that I'd not made some daft algebraic error or anything like that. Then I thought, well, if this is true, then there must be something wrong with my intuition, so my intuition needs fixing up. I needed to find a different way to think about the problem so that the answer did not seem so strange.

After some thought, I realized that the problem seems quite different if you think, for instance, about just a mile of the Earth's surface. If you lift the rope up by 16 cm over that mile, how much new rope do you need to add? Hardly any, obviously—the Earth is so big that it looks practically flat over that distance. In fact, the same would probably be true over 100 miles, or even more than that. So I found that I could imagine traveling around the Earth, lifting the rope up by 16 cm everywhere and adding just the rope I need as I go. I only need to add a tiny amount even over a long distance, so it now looks much more plausible that I might only need a meter to do the raising all the way around.

I found that this alternative point of view made me feel that the correct answer is not wholly unreasonable (if it doesn't work for you, can you come up with another alternative that does?). The point is that I came up with it because I made a deliberate effort to work out how to reconcile my intuitive thinking with the mathematically correct argument. I recognized that there must be a problem with my intuition, and I found a way to bring it into line with the mathematics, thus improving the connections between the two and increasing my likely power in using intuitive reasoning about related situations in future.

In this section and in the previous one, I've said a lot about counter-intuitive results. But I'm not saying that your intuition is going to be wrong about everything. Most of the time your mathematical experience will lead you to correct intuitive solutions (though, of course, there might be quite a bit of work to do to turn them into proofs). What I am saying is that mathematics does contain surprises like these, and that you should be alert for them. This is a bit like the message about definitions in Chapter 3, where I said that a definition probably means what you think it ought to mean in most cases, but that there might be "boundary" cases where it doesn't apply quite as you'd expect. This is like that. Occasionally your intuition will be wrong, and you should be prepared to put in a bit of effort to fix it up.

5.10 Will my whole major be proofs?

As I said at the beginning of this chapter, lots of the mathematics you meet during your major will be presented in the form of theorems and proofs. Some courses will be more about learning to perform standard calculations or to solve standard problems, but you will find that your professors present a lot of proofs and that they expect you to write a fair few too. Some people find this rather intimidating. They are accustomed to learning how to apply procedures, they see that every proof seems different, and they wonder how they will ever get to grips with all the presented proofs, never mind how they will learn to construct their own. This is a natural reaction, but in the remainder of Part 1 I hope to convince you that it isn't as hard as it might seem at first.

That said, when you start your major, you will probably have to increase your willingness to persevere in the face of difficult problems. My high-school teacher taught me a lot about this. More often than not, when I went to ask him about a question I was stuck with, he would look at me over the top of his glasses and say, "Oh, go away and think about it some more, Alcock, I'm sure you can do it if you try." Most of the time, he was right. More importantly, I learned that if I kept thinking for a bit longer, or had a break from the problem for a day then came back to it, often I could make progress. As a mathematics major you might find that you are regularly working on a single problem for half an hour or more, and that the length of time increases as you progress. That doesn't mean you're not good enough for mathematics at this level (see Chapter 13). In fact, if you're doing sensible things during the half hour, you will be doing a lot more than producing a single solution; you'll be revising existing knowledge, learning about the range of applicability of standard procedures, and so on.

You'll also be practising the strategies for getting started on a proof that we've discussed in this chapter: writing everything in terms of the relevant definitions, and thinking about example objects to see whether that gives you any insight. These are extremely general strategies, so they can be applied to almost any problem. We've also seen that one can often model the structure of a proof on the structure of a definition. In Chapter 6, we'll do a lot more than this, discussing strategies for proving different types of statement, and tricks that are often useful if you find yourself at an impasse.

SUMMARY

- Much advanced mathematics will be written in the form of theorems and proofs. This is not that different from high school mathematics, because much high school mathematics could be reformulated in these terms.
- It is often possible to get started on a proof by writing the premises and conclusions in terms of the relevant definitions.
- It is good practice to write a proof so that its structure reflects the structure of the relevant definition or theorem.
- Constructing a proof can require some creativity. Sometimes it is possible to get a good idea by examining different representations of relevant examples.
- It is possible to approach proof construction formally, thinking in terms of logical structures, or informally, thinking in terms of examples before translating into a formal argument. Each approach has advantages and disadvantages.
- Mathematicians sometimes prove obvious things in order to see how they fit together. Proving such things yourself allows you to practice the intellectual discipline needed for working from a small number of assumptions.
- Usually your intuition will be correct, but mathematics does contain some interesting counterintuitive results. When you meet one of these, you should use it as an opportunity to improve your intuitive understanding.
- Proofs sometimes appear difficult because they all look different. To master this aspect of mathematics, you will need to apply sensible strategies to get started, and to persevere.

FURTHER READING

For an introduction to developing formal proofs based on intuitive ideas, try:

- Katz, B. P. & Starbird, M. (2012). *Distilling Ideas: An Introduction to Mathematical Thinking*. Mathematical Association of America.

For more on constructing proofs appropriately for given logical statements, try:

- Allenby, R.B.J.T. (1997). *Numbers & Proofs*. Oxford: Butterworth Heinemann.
- Solow, D. (2005). *How to Read and Do Proofs*. Hoboken, NJ: John Wiley & Sons, Inc.
- Velleman, D.J. (2004). *How to Prove It: A Structured Approach*. Cambridge: Cambridge University Press.

For a question-based introduction to working with axiomatic systems, sequences, series, and functions, try:

- Burn, R.P. (1992). *Numbers and Functions: Steps into Analysis*. Cambridge: Cambridge University Press.

For examples and diagrams that provide insight into general results, try:

- Nelsen, R.B. (1993). *Proofs Without Words: Exercises in Visual Thinking*. Washington, DC: Mathematical Association of America.

For an introduction to mathematics education research results on students' understandings of proof, try:

- Reid, D.A. & Knipping, C. (2010). *Proof in Mathematics Education: Research, Learning and Teaching*. Rotterdam: Sense Publishers.

CHAPTER 6

Proof Types and Tricks

This chapter explains and illustrates some common proof structures. It describes aspects of proving that undergraduates sometimes find confusing, and explains why this confusion arises and how it can be overcome. It also points out some common tricks that you might see in proofs across a variety of courses, and gives guidance on how to approach novel proof tasks.

6.1 General proving strategies

In Chapter 4, I discussed two strategies for developing understanding of a theorem:

- Pay attention to the theorem's logical form, reading each sentence carefully.
- Think about examples of mathematical objects that satisfy the premises and consider how these relate to the conclusion.

In Chapter 5, I discussed the meaning of proof in advanced mathematics, some reasons for proving things carefully, and some particular proofs. When reading the proofs, you probably focused on the substance of each one—on the ideas used and on how they might generalize. However, I also mentioned some strategies that are generally useful when trying to construct a proof:

- Write the premises and conclusion in terms of the relevant definitions.
- Think about examples of mathematical objects to which the theorem applies, perhaps in terms of different representations.

In both cases, the first suggestion corresponds to a formal strategy and the second to a more informal, intuitive strategy. But it should be clear that these are really all just general strategies for mathematical problem solving. That is what we would expect, since constructing a proof is just a particular type of mathematical problem. It should also be clear that a student should never sit in front of a problem (proof or otherwise) and say "I don't know what to do." In particular, a student should not sit in front of a proof problem and think "I can't do this because I haven't been shown how." Mathematics majors are expected to show some initiative, and trying any of these strategies will be better than doing nothing.

However, when learning to construct and understand proofs, it also helps to be aware that there are some standard proof types and standard tricks that will often appear and will often be useful. If you are alert to these, you should find that you can see relationships between mathematical arguments across a variety of courses and that, when you get stuck, you can think up ideas that might allow you to make progress. This chapter is about these proof types and tricks.

6.2 Direct proof

The first standard proof type is known as *direct proof*. In a direct proof we start by assuming that the premise(s) hold and move, via a sequence of valid manipulations or logical deductions, to the desired conclusion. Most of the proofs in this book so far have been like this. Here are some theorems for which we've already studied a direct proof:

Theorem: If n is an even number, then any integer multiple of n is even.

Theorem: If $x^2 - 20x + 10 = 0$ then $x = 10 + 3\sqrt{10}$ or $x = 10 - 3\sqrt{10}$.

Theorem: $(2, 5)$ is an open set.

Theorem: For every $\theta \in \mathbb{R}$, $\cos(3\theta) = 4\cos^3\theta - 3\cos\theta$.

Theorem: If $ax^2 + bx + c = 0$, then $x = \dfrac{-b \pm \sqrt{b^2 - 4ac}}{2a}$.

Theorem: Suppose that f and g are both differentiable. Then

$$\frac{\mathrm{d}}{\mathrm{d}x}(f + g) = \frac{\mathrm{d}f}{\mathrm{d}x} + \frac{\mathrm{d}g}{\mathrm{d}x}.$$

Theorem: If l, m, and n are consecutive integers, then the product lmn is divisible by 6.

Theorem: If f is an even function, then $\dfrac{\mathrm{d}f}{\mathrm{d}x}$ is an odd function.

Stop and think for a moment here. Can you write out proofs of these statements without looking? If you can't do it immediately, can you remember the gist of how we proceeded in each case and reconstruct the rest? If you give yourself a minute for each one, I bet you can remember more than you would initally have thought. Students often have too little faith in their own ability to recall mathematical ideas and reconstitute arguments around them.

One important thing to note is that *direct* describes the eventual proof we write, it does **not** necessarily describe the process of constructing the proof. You might be able to write down the premises and just follow your nose to a proof, but it is more likely that you'll have to try out some of the things I suggested when discussing these theorems: write everything in terms of definitions, think about some examples, maybe draw a diagram, and so on. You should then, however, work out how to write your final proof in a way that makes its logical structure clear for a reader. It is probably a good idea to treat this writing as a separate task, one that is worthy of your attention over and above simply getting to an answer or a proof. I'll have lots more to say about this in Chapter 8.

A second thing to note is that direct proofs can have somewhat more complicated structures within them. The most obvious such structure occurs in a *proof by cases*, which means what it sounds like it means: we divide up the cases we're dealing with into sensible groups and work with each one separately within the main proof. Consider, for instance, the piecewise-defined increasing function we looked at in Chapter 3:

$$f(x) = \begin{cases} x+1 & \text{if } x < 0 \\ 1 & \text{if } 0 \le x \le 1 \\ x & \text{if } x > 1 \end{cases}$$

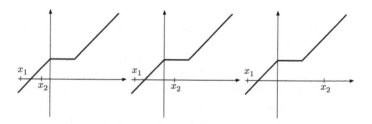

Looking at the diagram, we were able to imagine that we can let x_1 and x_2 slide around but, as long as we keep $x_1 < x_2$, we always have $f(x_1) \le f(x_2)$. In a proof, though, we can't talk about "sliding things around," we have to be more precise. In fact, x_1 and x_2 can only be in one of six different positions relative to the sections of the function, three of which are shown here (what are the others like?):

In a proof, we can deal with each of these situations as a separate case. For these three cases, a proof might look like that given below. What would the remaining cases be and what would you write?

Theorem:

$$f : \mathbb{R} \to \mathbb{R} \text{ given by } f(x) = \begin{cases} x+1 & \text{if } x < 0 \\ 1 & \text{if } 0 \le x \le 1 \\ x & \text{if } x > 1 \end{cases}$$

is increasing.

Proof:

Case 1:

Suppose that $x_1 < x_2 < 0$.
Then $f(x_1) = x_1 + 1 < x_2 + 1 = f(x_2)$.

Case 2:

Suppose that $x_1 < 0 \leq x_2 \leq 1$.
Then $f(x_1) = x_1 + 1 < 0 + 1 = 1 = f(x_2)$.

Case 3:

Suppose that $x_1 < 0$ and $x_2 > 1$.
Then $f(x_1) = x_1 + 1 < 0 + 1 = 1 < x_2 = f(x_2)$.

. . .

For each case I have used chains of equalities and inequalities with $f(x_1)$ on one end and $f(x_2)$ on the other end; you should make sure that you can see exactly why each equality or inequality in the chain is valid. I like chains of equalities and inequalities like these because they are clearly linked to the conclusion of the theorem (they show that $f(x_1) < f(x_2)$ in each case) and because I think they're elegant. If you prefer, you could write down what $f(x_1)$ is and what $f(x_2)$ is separately and then compare them. You'll end up writing more, though.

So when should you consider a proof by cases? Sometimes you will find that you have to. In this illustration, for instance, we don't have much choice; the $f(x)$ values are different depending upon the positions of x_1 and x_2, so we have to handle these cases separately. In other situations, it might not be necessary to use a proof by cases, but it might be convenient anyway because there is some sort of natural split, perhaps between positive and negative numbers, or between odd and even ones. Finally, it might be worth starting a proof by cases if you think that you can construct an argument for some of the objects to which the theorem applies but not for others. A start is better than nothing and, once you've got an argument written down for one case, you might find that reflecting on it gives you ideas about how to continue.

One final tip is that, if you have produced a proof by cases, or if you're looking at one produced by someone else, it might be a good idea to ask whether the number of cases could be reduced. Perhaps, for instance, you've separately treated the three cases $x < 0$, $x = 0$, and $x > 0$, but you could rewrite with just one argument for $x < 0$ and one for $x \geq 0$. You don't have to do this, of course—if your proof is valid as it is, that's fine. But remember that mathematicians also value elegance, and brevity contributes to elegance so it's a worthwhile aim.

6.3 Proof by contradiction

The second type of proof I want to talk about is *proof by contradiction*. This is a type of indirect proof, so called because we do not proceed directly from the premises to the conclusion. Instead, we make a temporary assumption that our desired conclusion (or some part of it) is false, and we show that this leads us to a contradiction. From this we can deduce that the temporary assumption must have been wrong, and thus that the desired conclusion is true. This sounds rather convoluted, but you're accustomed to making this kind of argument informally in everyday life. Here's a simple case:

Your friend: Dan was at home in Hereford all weekend.

You: No he wasn't, I saw him in Coventry on Saturday afternoon.

Here you are implicitly using a proof by contradiction. The temporary assumption is that Dan was in Hereford. Your argument says that if we make that assumption, then we can deduce that he wasn't in Coventry (if you like, this uses the "theorem" that people can't be in two places at the same time). But this is contradicted by the fact that you saw him in Coventry. So the assumption that he was in Hereford must have been wrong.

Next, we will look at a mathematical example, which involves the definition of *rational number*. We introduced this informally in Chapter 2, along with the notation $\mathbb{Q}$ to denote the set of all rational numbers. More formally we have:

Definition: $x \in \mathbb{Q}$ if and only if $\exists\, p, q \in \mathbb{Z}$ (with $q \neq 0$) such that $x = p/q$.

The theorem and proof below use the symbol $\notin$ to mean "is not an element of," and in this case $y \notin \mathbb{Q}$ means that y is an *irrational number*. The theorem and proof both implicitly assume that all the numbers we are working with are real (this is common in early work with rational and irrational numbers). As with any theorem and proof, you should read everything carefully, checking that you understand what is going on in each step.

Theorem: If $x \in \mathbb{Q}$ and $y \notin \mathbb{Q}$ then $x + y \notin \mathbb{Q}$.

> **Proof:** Let $x \in \mathbb{Q}$, so $\exists\, p, q \in \mathbb{Z}$ (with $q \neq 0$) such that $x = p/q$.
>
> Let $y \notin \mathbb{Q}$.
>
> Suppose for contradiction that $x + y \in \mathbb{Q}$.
>
> This means that $\exists\, r, s \in \mathbb{Z}$ (with $s \neq 0$) such that $x + y = \dfrac{r}{s}$.
>
> But then $y = \dfrac{r}{s} - x = \dfrac{r}{s} - \dfrac{p}{q} = \dfrac{rq - ps}{sq}$.
>
> Now $rq - ps \in \mathbb{Z}$ and $sq \in \mathbb{Z}$ because $p, q, r, s \in \mathbb{Z}$.
>
> Also $sq \neq 0$ because $q \neq 0$ and $s \neq 0$.
>
> So $y \in \mathbb{Q}$.
>
> But this contradicts the theorem premise.
>
> So it must be the case that $x + y \notin \mathbb{Q}$.

In this proof, the temporary assumption is this one:

> Suppose for contradiction that $x + y \in \mathbb{Q}$.

Making that temporary assumption leads us, by some sensible use of definitions and algebra, to this line:

> So $y \in \mathbb{Q}$.

This (as stated) contradicts the theorem premise, so it allows us to conclude that our temporary assumption must have been wrong, like this:

> So it must be the case that $x + y \notin \mathbb{Q}$.

It should be clear that in order to properly understand a proof like this, you have to do more than you might have had to do in earlier mathematics. In high school, most of your mathematical reading will have involved checking some algebra. Here, you have to be more sophisticated. You certainly should check to make sure that you can see how the algebra works and that there are no mistakes. But that's not really where the action is in a proof like this. To fully understand it, you need to understand its global structure. You need to be able to identify what assumptions are made where, identify where the contradiction arises and what exactly is being contradicted, and understand how it all fits together to prove that the theorem is true.

You will see proofs by contradiction in a variety of courses, and one of the first will probably be a proof that $\sqrt{2}$ is irrational. If you think about this properly, you might be able to see that it is very far from obvious. There are *lots* of rational numbers. Between 1 and 2, for instance, we could start listing $\frac{3}{2}, \frac{4}{3}, \frac{5}{3}, \ldots$, and of course there are also things like $\frac{11}{7}$. Hence, when we say that $\sqrt{2}$ is irrational, we are making a very strong claim: from all these possibilities, there is *not a single one* that is exactly equal to $\sqrt{2}$. That includes all the very non-obvious examples like 29 365 930 375/16 406 749 305. We might be able to get a good approximation by considering such examples, but we will never get the exact number. You can imagine that people might have been pretty upset when they first found this out.

Now, I'm not going to show you a proof that $\sqrt{2}$ is irrational. It's a classic proof, you'll certainly see it during a mathematics major, and I don't want to steal your professor's thunder. But I want to use the idea to discuss the use of proof by contradiction, and to revisit some points from earlier chapters.

First, notice that although the claim that $\sqrt{2}$ is irrational *sounds* like a statement about a particular number, it is really a statement about infinitely many things. It can be rephrased as:

There is no number of the form p/q that is equal to $\sqrt{2}$.

This, in turn, can be rephrased in our standard theorem form as:

If $x \in \mathbb{Q}$, then $x \neq \sqrt{2}$.

As I said at the end of Chapter 4, it's important that you learn to recognize when statements are and are not logically equivalent. Make sure you can see the equivalence in this case.

A proof by contradiction for this theorem would start by assuming that there *is* a rational number equal to $\sqrt{2}$, and show that this leads to a contradiction. Look out for this structure when you see a proof and, as ever, think about how the proof would generalize. I say this to reiterate a point from Chapter 1: it is dangerous to copy a mathematical argument without thinking carefully about whether each step really holds in a new situation. One mathematics professor I know likes to demonstrate that $\sqrt{2}$ is irrational then set a problem in which she asks students to prove that $\sqrt{3}$ is irrational, then that $\sqrt{4}$ is irrational. What do you notice about the last one? Yep, whatever the proof does, there must be a step in it that does not work for $\sqrt{4}$. But, my friend tells me, lots of students merrily copy out the proof for $\sqrt{4}$ nonetheless. You don't want to be one of those students, so stop and think.

A final point I want to make is that mathematicians behave in a slightly out-of-character way when introducing this proof. As I said in Chapter 5, mathematicians usually like to work with general versions of statements. You would think, in this case, that they wouldn't bother working with $\sqrt{2}$, and would instead go straight to proving the true and more general theorem:

Theorem: If k is prime then $\sqrt{k}$ is irrational.

They don't, usually. They usually start with the $\sqrt{2}$ case. I think this is partly because of the geometric roots of the problem, and partly because it usually comes up early in a student's major and mathematicians think it will be easier to start with a specific case. That's probably true but, as ever, you should think about generalizations for yourself.

So, when should you try a proof by contradiction? Your professors will use this method in many different situations, so you should get a sense of this as you go along. But sometimes the phrasing of a question or theorem might lead you to think of proof by contradiction quite naturally. For instance, if a question says "Prove that such-and-such an object does not exist," you might make progress by assuming that one does exist and showing that this leads to a contradiction. It might also be useful if you are looking at a theorem in the standard form "If *something* then *something else*." Here you might make progress by assuming that the premises hold but the conclusion does not, especially if you consider the theorem so obvious that you wouldn't otherwise know how to begin. Ask yourself: "What would go wrong if the conclusion did *not* hold?" Sometimes that unlocks the problem.

6.4 Proof by induction

The final standard proof type I want to discuss is *proof by induction*. Depending on your previous experience, you may have met this already. If so, you'll have used it to prove things like this:

$$\sum_{i=1}^{n} i^2 = \frac{n(n+1)(2n+1)}{6}.$$

If not, you might not have seen this notation so here is a quick explanation. The symbol Σ is called "sigma" and is a Greek upper case letter S, used here to denote a sum from $i = 1$ to $i = n$. Written out, the left-hand side of the above means

$$\sum_{i=1}^{n} i^2 = 1^2 + 2^2 + 3^3 + \cdots + n^2.$$

What we've done here is substitute $i = 1$, then $i = 2$, then $i = 3$, and so on, up to $i = n$, where we stop.

So our original expression actually captures infinitely many propositions:

$$P(1) \qquad\qquad 1^2 \quad = \quad \frac{1(1+1)(2+1)}{6}$$

$$P(2) \qquad\qquad 1^2 + 2^2 \quad = \quad \frac{2(2+1)(4+1)}{6}$$

$$P(3) \qquad 1^2 + 2^2 + 3^2 \quad = \quad \frac{3(3+1)(6+1)}{6}$$

$$P(4) \quad 1^2 + 2^2 + 3^2 + 4^2 \quad = \quad \frac{4(4+1)(8+1)}{6}, \text{ and so on.}$$

Having a theorem that captures infinitely many cases isn't unusual—many of the other theorems we've looked at do the same. The difference here is that the form of the statement allows us to put the propositions in an ordered list $P(1), P(2), P(3), P(4), \ldots$.

Proof by induction works as follows. First we prove $P(1)$. This is often easy. Then we do something clever. We don't try to prove any of the other propositions directly. Instead, we take a general number k and prove that if $P(k)$ is true, then $P(k + 1)$ must be true too. This gives us $P(1) \Rightarrow P(2)$ and, since we've already proved $P(1)$, we can conclude that $P(2)$ is true as well. It also gives us $P(2) \Rightarrow P(3)$, so we can conclude that $P(3)$ is true as well. You get the idea. We have made an infinite chain of propositions, which are all true because the first one is true and the implications are all true:

$$P(1) \Rightarrow P(2) \Rightarrow P(3) \Rightarrow P(4) \Rightarrow \cdots .$$

Proof by induction is one of those ideas that students usually find intuitively straightforward when it is explained in the abstract. However, they often find it difficult to use in any particular case, so we will look closely at the example we started with. In that example, it is easy to prove that $P(1)$ is true:

$$1^2 = 1 \text{ and } \frac{1(1+1)(2+1)}{6} = 1.$$

Proving that the implication $P(k) \Rightarrow P(k + 1)$ is a bit harder. We would like to assume that $P(k)$ is true and use this to prove that $P(k + 1)$ is true. I would start doing some rough work at this point, writing something like this:

Will assume $P(k)$, which means $\displaystyle\sum_{i=1}^{k} i^2 = \frac{k(k+1)(2k+1)}{6}$.

Want to prove $P(k + 1)$, which means

$$\sum_{i=1}^{k+1} i^2 = \frac{(k+1)((k+1)+1)(2(k+1)+1)}{6},$$

which, by rewriting the left-hand side, means we want

$$\left(\sum_{i=1}^{k} i^2\right) + (k+1)^2 = \frac{(k+1)((k+1)+1)(2(k+1)+1)}{6},$$

which, by the assumption about $P(k)$, means we want

$$\frac{k(k+1)(2k+1)}{6} + (k+1)^2 = \frac{(k+1)((k+1)+1)(2(k+1)+1)}{6}.$$

Then I've just got some algebra to do to show that the last two things are, in fact, equal (you might like to try it).

This, however, is definitely a situation in which the way you think about constructing a proof is not necessarily the same as the way you should write it out. The thinking above is completely logical, but presenting it like that wouldn't work very well because it doesn't match the structure of what we are trying to prove. When we write out a proof that $P(k) \Rightarrow P(k+1)$, we really want our proof to start with a clear assumption of $P(k)$ and proceed through some nice, tidy deductions to $P(k+1)$.

For proofs by induction, I favor a layout that makes that structure very clear. I would write something like this:

Theorem: $\forall n \in \mathbb{N}, \displaystyle\sum_{i=1}^{n} i^2 = \frac{n(n+1)(2n+1)}{6}.$

Proof (by induction):

Let $P(n)$ be the statement that $\displaystyle\sum_{i=1}^{n} i^2 = \frac{n(n+1)(2n+1)}{6}.$

Note that $1^2 = 1 = \dfrac{1(1+1)(2+1)}{6}$ so $P(1)$ is true.

Now let $k \in \mathbb{N}$ be arbitrary and assume that $P(k)$ is true, i.e. that

$$\sum_{i=1}^{k} i^2 = \frac{k(k+1)(2k+1)}{6}.$$

Then

$$\sum_{i=1}^{k+1} i^2 = \left(\sum_{i=1}^{k} i^2 \right) + (k+1)^2$$

$$= \frac{k(k+1)(2k+1)}{6} + (k+1)^2 \text{ by the assumption}$$

$$= \frac{k(k+1)(2k+1) + 6(k+1)^2}{6}$$

$$= \frac{(k+1)(k(2k+1) + 6(k+1))}{6}$$

$$= \frac{(k+1)(2k^2 + 7k + 6)}{6}$$

$$= \frac{(k+1)(k+2)(2k+3)}{6}$$

$$= \frac{(k+1)((k+1)+1)(2(k+1)+1)}{6}.$$

So $\forall k \in \mathbb{N}, P(k) \Rightarrow P(k+1)$.

Hence, by mathematical induction, $P(n)$ is true $\forall n \in \mathbb{N}$.

There are a couple of things to notice about this. First, the proof contains only a few words, but these help to make the structure clear. Second, all the algebra is in a single chain of equalities that starts with the left-hand side of the statement of $P(k+1)$ and ends with the right-hand side. You might like to think about why it makes sense to do the manipulations in this order, given that we know what we're going for. You might also like to think about why the last expression in the chain is not necessary but might be useful for a reader who wants to link the proof back to the theorem.

In general, you will see some variation in the way that people write proofs by induction. Some people frame the whole thing in terms of a "successor function," where the successor of 1 is 2, the successor of 2 is 3, and so on. Some people explicitly write "base case" when they are proving $P(1)$ and "induction step" when they are proving that $P(k) \Rightarrow P(k+1)$. Some people don't use names like $P(n)$ and instead write out the exact statement each time. That's fine, although I find that explicit naming can keep the structure clear, especially when you're trying to work out what you want to prove in the first place. Some people don't introduce a k at all, but call their variable n throughout. I find that doing so makes students a bit more prone to error, as it means that n is being used in two different ways, which raises the risk of confusion.

Confusion does tend to arise with proof by induction because there are lots of things to think about. I find that students are confused most often by the point at which we write, "Assume that $P(k)$ is true." Students often read this and think, "But that's what we want to prove, how come we're allowed to assume it?" In fact, at that stage in the proof, we are **not** proving that $P(k)$ is true, we are proving that $P(k) \Rightarrow P(k+1)$. Make sure you can see the difference. Also, students sometimes confuse themselves, usually because they have allowed ambiguities to creep into their writing by using the word "it" in phrases like "so it is true for n." There are many possible candidates for the meaning of "it" in a typical proof by induction, so you should be more specific. Writing "So $P(n)$ is true" is one way to do that. (Notice that the proof above does not include the word "it"—we are very specific about what we have deduced at each stage.)

So when should you use proof by induction? In some cases this will be obvious, because you are likely to have a section about it in at least one course. Also you will come across cases in which you want to prove that something is true for all $n \in \mathbb{N}$, which is a giveaway that induction is worth a try. Be aware, however, that problems for which induction is useful can vary quite a bit. First, there is no particular reason for a proof by induction to start at $n = 1$. You might be asked to prove that something is true for every $n \in \mathbb{N}$ such that $n > 4$, for instance. In that case, you can just make $P(5)$ your base case and proceed as before, except that at some point, perhaps in the induction step, you will find that you need $n > 4$ to justify some manipulation you want to make. Second, while some of the first problems you meet will involve working with a sum, proof by induction is useful for many other types of problem. All we really need is a situation in which we have infinitely many statements that can be listed in the order of the natural numbers, which can happen in all sorts of ways. For instance, consider these tasks:

Prove that for every natural number $n > 10$, $2^n > n^3$.

Prove that for every $n \in \mathbb{N}$, $5^{3n} + 2^{n+1}$ is divisible by 3.

For the first task, we would write

Let $P(n)$ be the statement that $2^n > n^3$.

Then we would prove directly that $P(11)$ is true, i.e. that $2^{11} > 11^3$. Then we would work out how to prove that if $k > 10$ and $2^k > k^3$, then $2^{k+1} > (k+1)^3$.

For the second task, we would write

Let $P(n)$ be the statement that $5^{3n} + 2^{n+1}$ is divisible by 3.

Then we would prove directly that $P(1)$ is true, i.e. that $5^3 + 2^2$ is divisible by 3. Then we would work out how to prove that if $5^{3k} + 2^{k+1}$ is divisible by 3, then $5^{3(k+1)} + 2^{(k+1)+1}$ is divisible by 3. Notice, in this case, that the statement $P(k)$ is **not** just "$5^{3k} + 2^{k+1}$." In fact, $5^{3k} + 2^{k+1}$ isn't a statement at all—we couldn't prove it because it's just an expression (for any particular k it is a number; you can't "prove" a number, and it makes no sense to say that one number implies another). The statement is "$5^{3n} + 2^{n+1}$ is divisible by 3."

In my experience, starting out with a clear statement of $P(n)$ often makes the difference between success and failure in constructing a proof by induction, especially when dealing with a new type of problem. This isn't that surprising, of course—before you start any problem, you should always make sure that you are clear about what you are trying to do. In any case, as I said in Chapter 1, you won't always be told what method to use, so you should be on the lookout for less familiar cases like these, and you should train yourself to notice when proof by induction might be useful.

6.5 Uniqueness proofs

The standard proof types I've talked about so far are all very general. The terms *direct proof*, *proof by contradiction*, and *proof by induction* describe the structure of a whole proof, and they will appear across many different undergraduate courses. Next, I want to discuss a couple of tricks that are also quite general but that do not really constitute types of proof, as such. These tricks are logically sound so they could, in principle, be used anywhere in mathematics, but you'll probably find that you see them more in some subjects than in others.

One of these tricks is a way to prove the uniqueness of some object. That is, to prove that there is exactly one object that has a certain property. As a student, you might find that it is obvious to you that there must be only one of this particular thing. In those cases you'll have to remember that we're not always proving things because we doubt that they're true; sometimes we're proving them so that we can see how everything fits into a bigger theory. In this sense, you might think of uniqueness proofs as tying up loose ends.

The standard trick when trying to prove that something is unique is to start by assuming that there are two different ones and prove that, in fact, they must be the same. Below is an illustration. It involves the idea of an *additive identity*, which is defined as follows:

Definition: Let S be a set. We say $k \in S$ is an *additive identity* for S if and only if $\forall s \in S, s + k = k + s = s$.

Remember to read this definition carefully. Can you think of a number that would be an additive identity for the set $S = \mathbb{Z}$? Can you think of more than one? The theorem and proof below captures this, and illustrates the trick.

Theorem: The additive identity for the integers is unique.

Proof: Suppose that there are two integers 0 and $0'$ that are both additive identities for the integers.

Then, by definition,
(1) $\forall x \in \mathbb{Z}, x + 0 = x$ and (2) $\forall x \in \mathbb{Z}, 0' + x = x$.

In particular,

$$0' = 0' + 0 \text{ using (1)}$$

$$= 0 \text{ using (2).}$$

So $0' = 0$, so the additive identity 0 is unique.

Some people find this type of argument a bit strange when they first meet it, because it involves introducing something you know is going to be the same as something else, and temporarily pretending that you don't know that. However, I think it is rather elegant. It is brief, and it has a nice symmetry to it. It is also quite straightforwardly generalizable. Can you see how it could be adapted to prove that the multiplicative identity 1 for the integers is also unique? You will find that such arguments crop up in contexts that involve axiomatic structures, such as Abstract Algebra. But they also crop up in other contexts. One can, for instance, prove that a function cannot have two different limits as $x \to \infty$ by assuming that there are two limits, giving them different names, then proving that they must be equal. You'll likely encounter that type of argument in a course called Advanced Calculus or Analysis.

Other common tricks share with this one the property that they can be usefully applied when a result seems obvious. For instance, you might see a professor prove that two numbers a and b must be equal to each other by proving first that $a \leq b$ then that $a \geq b$. Or proving that two sets A and B must be equal to each other by proving first that $A \subseteq B$ then that $B \subseteq A$. As with all proof types and tricks, looking out for this kind of regularity should help you to get to grips with your whole major.

6.6 Adding and subtracting the same thing

Another common trick is to add and subtract the same thing to make it possible to split up an expression into parts that are easier to work with. A good illustration of this can be seen in the standard proof of the product rule for differentiation. Before you read this, it might be a good idea to turn back to Chapter 5 and look again at the proof of the sum rule, so you can appreciate the contrast. Here is the product rule:

Theorem: Suppose that f and g are both differentiable. Then

$$\frac{\mathrm{d}}{\mathrm{d}x}(fg) = f(x)\frac{\mathrm{d}g}{\mathrm{d}x} + g(x)\frac{\mathrm{d}f}{\mathrm{d}x}.$$

As usual, we can understand this by thinking about what the premises and conclusion mean in terms of the definition. Here the premise is that f and g are both differentiable, so their derivatives are defined according to the definition as

$$\frac{\mathrm{d}f}{\mathrm{d}x} = \lim_{h \to 0} \frac{f(x+h) - f(x)}{h} \text{ and } \frac{\mathrm{d}g}{\mathrm{d}x} = \lim_{h \to 0} \frac{g(x+h) - g(x)}{h}.$$

The conclusion, written in terms of definitions, says that

$$\lim_{h \to 0} \frac{(fg)(x+h) - (fg)(x)}{h}$$

$$= f(x) \lim_{h \to 0} \frac{g(x+h) - g(x)}{h} + g(x) \lim_{h \to 0} \frac{f(x+h) - f(x)}{h}.$$

Proving this will not be as straightforward as it was in the sum rule case. We can't just "split up" the left-hand side of the equality in the conclusion, even if we write

$$(fg)(x+h) - (fg)(x) = f(x+h)g(x+h) - f(x)g(x).$$

We can, however, do some useful splitting up if we first add and subtract the expression $f(x+h)g(x)$, and write

$$(fg)(x+h) - (fg)(x)$$
$$= f(x+h)g(x+h) - f(x+h)g(x) + f(x+h)g(x) - f(x)g(x).$$

Here is how:

Proof: Suppose that f and g are differentiable. Then

$$\frac{\mathrm{d}}{\mathrm{d}x}(fg) = \lim_{h \to 0} \frac{(fg)(x+h) - (fg)(x)}{h}$$

$$= \lim_{h \to 0} \frac{f(x+h)g(x+h) - f(x)g(x)}{h}$$

$$= \lim_{h \to 0} \frac{f(x+h)g(x+h) - f(x+h)g(x) + f(x+h)g(x) - f(x)g(x)}{h}$$

$$= \lim_{h \to 0} \frac{f(x+h)\left(g(x+h) - g(x)\right) + g(x)\left(f(x+h) - f(x)\right)}{h}$$

$$= \lim_{h \to 0} \frac{f(x+h)\left(g(x+h) - g(x)\right)}{h} + \lim_{h \to 0} \frac{g(x)\left(f(x+h) - f(x)\right)}{h}$$

$$= \lim_{h \to 0} f(x+h) \lim_{h \to 0} \frac{g(x+h) - g(x)}{h} + \lim_{h \to 0} g(x) \lim_{h \to 0} \frac{f(x+h) - f(x)}{h}$$

$$= f(x)\frac{dg}{dx} + g(x)\frac{df}{dx}, \text{ as required!}^1$$

This trick is somewhat like the previous one. In the uniqueness case, we introduced something that we knew would turn out to be the same as something else. In this case, we introduce something that is equal to zero and that eventually gets incorporated in a convenient way into our calculations. Another similar trick involves subtracting something from an expression to get something that is easier to work with, then adding it back on later.

You will see numerous variations on such tricks in proofs presented by your professors and I hope that you will appreciate that they can be very mathematically elegant. However, if you are like most people, seeing such elegance in proofs produced by others might make you a bit concerned because you do not feel you could invent such elegant strategies yourself. I'll say something about that at the end of this chapter, but first a word about trying things out.

6.7 Trying things out

At the beginning of this chapter I said that a student should never sit in front of a problem and think "I don't know what to do." There are always things to try and, to be a good mathematician, you must be willing to try them. In fact, you must be willing to try things that turn out not to work. In my experience, students are sometimes unwilling to do this, for three main reasons.

First, some students dislike the insecurity of not knowing exactly what to do. It makes them nervous. They want to know in advance what is going to work, and sometimes they ask for a teacher's assurance about this ("Is this the right way to do it?"). The problem with seeking such assurance all the time is that you never find out what you could have done if you'd had a go, which means that you never get any more confident, which

---

[1] The last step in the chain of equalities is valid because as $h \to 0$ we have $x + h \to x$, so $f(x + h) \to f(x)$. Also, $\lim_{h \to 0} g(x) = g(x)$ automatically because $g(x)$ does not depend on h so it does not change as $h \to 0$. Finally, the whole proof relies on theorems about sums and products of limits; your professor might prove these theorems and highlight where they are used.

means that you end up in a vicious circle, having to ask for support all the time.

Second, some students do not want to waste time. I understand this—obviously no-one wants to spend ages on one thing, especially when there are so many interesting things to do at college. But it is a big mistake to think that trying something that turns out not to work is a waste of time. Time spent learning is never wasted. If you try a method that doesn't work then, provided you are thoughtful about it, you learn *why* it doesn't work, which means you know something new about the applicability of the method. And you might gain some insight about the problem so that you have a better idea about what to try next. Of course, it is a mistake to keep plugging away at a method that clearly isn't working—research shows that good problem solvers stop frequently to re-evaluate whether their current approach seems to be getting them anywhere. But it's an even bigger mistake not to start.

Third, some students do not want to mess up their paper. They want to know that once they begin writing, they will be able to carry on writing and arrive at a nice, neat, correct solution. If this applies to you, then I'm afraid you'll have to get over it. Real mathematical thinking is not tidy. It is full of false starts and partial attempts and realizations that what does not seem to be working just here would, in fact, form a useful part of a solution if put together with something that failed ten minutes ago, or yesterday, or last week. It is very important to embrace this if you want to keep improving as a mathematical problem solver. You need to get partial solution attempts on paper for the simple, practical reason that your brain cannot handle many things at once. You have an enormous amount of knowledge stored in what is known as your long-term memory, but your working memory, where you actually do the new thinking, has a seriously limited capacity. It will not be big enough to hold all the information about a complicated mathematical problem while simultaneously working out how to solve it. When you write down representations or definitions or theorems or calculations that might help you solve a problem or construct a proof, you are using the paper to supplement your cognitive powers by, in effect, extending the capacity of your working memory. So don't be worried about writing things that are wrong or that turn out not to be useful. You can always write up a neat version of your solution or proof later.

6.8 "I would never have thought of that"

Your professors will explain proofs that are long and logically complicated. They will also explain proofs that rely on some really clever insight. Sometimes they will explain proofs that are long and logically complicated *and* that rely on some really clever insight. This tends to worry students. They think, "Well, okay, I can see how that works, but *I would never have thought of it*." This makes them wonder whether they're good enough at mathematics. But you shouldn't worry, because you're not supposed to be able to reinvent the whole of modern mathematics by having all the original ideas yourself. Even a mathematics PhD student wouldn't be expected to have many totally original insights. As an undergraduate, when faced with a proof like this, your job is to appreciate the clever insight, to understand why it works, to think about how modifications of it might work in slightly different circumstances, and to relate it to ideas used elsewhere in the course or in your major. To reassure you further, here is a list of things that you *will* be expected to do.

First, you will be expected to do routine mathematical calculations much like those you have seen in lower-level mathematics. As I said in Chapter 1, you should be prepared for these calculations to be longer and more involved than those you have experienced before, and you should be prepared to have to adapt the calculation procedure if a step in it is not valid for a new case.

Second, you will be expected to adapt proofs that you have seen to closely related cases. For instance:

- Having seen a proof that $\sqrt{2}$ is irrational, you might be expected to prove that $\sqrt{3}$ is irrational.
- Having seen a proof that the function f given by $f(x) = 3x$ is continuous, you might be expected to prove that the function g given by $g(x) = -10x$ is continuous.

In such cases, you will often be able to treat the proof you have seen like a template, and change some numbers appropriately. However, to reiterate one of the main points from Chapter 1, you shouldn't do this thoughtlessly—you should make sure that each step in the proof really does work for the new case, and be ready to make minor adjustments if it doesn't. It is important to be careful in cases where some number might be zero, for instance, or when dividing both sides of an inequality by a number that might be negative.

Third, you will be expected to adapt proofs you have seen to cases that are related, but not so closely. For instance:

- Having seen a proof that $\frac{d}{dx}(f+g) = \frac{df}{dx} + \frac{dg}{dx}$, you might be expected to prove that $\frac{d}{dx}(cf) = c\frac{df}{dx}$.
- Having seen a proof that an increasing sequence that is bounded above has a limit, you might be expected to prove that a decreasing sequence that is bounded below has a limit.

In cases like these, the proof you have seen will certainly be helpful, but you will not be able to treat it like a template. You might be able to construct a proof that is very similar in its basic structure, but you will have to think further to work out exactly what needs changing.

Fourth, you will be expected to show that definitions are satisfied. You will sometimes be expected to do this with definitions you have not seen before, if your professor thinks that they are sufficiently straightforward. I talked about this in Chapter 3.

Fifth, you will be expected to construct proofs of theorems for which you haven't seen a closely related model. As I've said, the biggest mistake you could make here would be to sit around thinking "We haven't been shown how to do this." No-one will ask you to do things that are completely beyond you, and this chapter has been about things you might try.

Sixth and finally, on an exam you might be asked to state and prove some of the more challenging theorems from the course. Sometimes a question might lead you through such a proof in steps, or might offer a fairly big hint to remind you of a key idea or a useful trick. Sometimes it might just ask outright, which means you'll have to be able to remember the key ideas or useful tricks for yourself and reconstruct the rest. To do so you will need to have effectively read and understood the material in your lecture notes. How to go about this is the subject of the next chapter.

SUMMARY

- A direct proof proceeds from the premises of a theorem to the conclusion by a sequence of valid manipulations or logical deductions.
- A proof by cases works by splitting up the mathematical objects under consideration and dealing separately with different sets of them.
- A proof by contradiction involves making a temporary assumption that the conclusion or some part of it is not true, and showing that this leads to a contradiction so that the temporary assumption must have been false.
- A proof by induction is often used to prove that a proposition $P(n)$ holds for every natural number n. It involves proving a base case, often $P(1)$, then proving the induction step $P(k) \Rightarrow P(k + 1)$.
- A common strategy for proving that a mathematical object is unique is to assume that there are two different ones, then show that they must be equal.
- Some proofs use a trick such as adding and subtracting the same thing in order to get an expression that is easier to work with.
- A mathematics student must be willing to try things out; you will not always know in advance whether something is going to work.
- When you see a proof that involves a clever insight, you should not worry that you wouldn't have thought of it; you should think about why it works, how it might be adapted, and how it relates to other ideas you have seen.

FURTHER READING

For more on types of proof and how to construct them, try:

- Solow, D. (2005). *How to Read and Do Proofs*. Hoboken, NJ: John Wiley.
- Velleman, D.J. (2004). *How to Prove It: A Structured Approach*. Cambridge: Cambridge University Press.
- Allenby, R.B.J.T. (1997). *Numbers & Proofs*. Oxford: Butterworth Heinemann.
- Houston, K. (2009). *How to Think Like a Mathematician*. Cambridge: Cambridge University Press.
- Vivaldi, F. (2011). *Mathematical Writing: An Undergraduate Course*. Online at http://www.maths.qml.ac.uk/~fv/books/mw/mwbook.pdf.

CHAPTER 7

Reading Mathematics

This chapter explains the importance of developing skills in learning mathematics independently from written information such as lecture notes. It explains strategies that might be useful in reading for understanding, for synthesis, and for memory. It also discusses how to build on these strategies when preparing for exams.

7.1 Independent reading

How much mathematics have you learned by reading something on your own, with no verbal explanation from a teacher? The answer might be "quite a lot," especially if your high school teacher often asked you to study something before a class or if, say, you took an advanced placement course that involved some independent study. But it might be "not very much at all, actually." Perhaps your high school teacher usually explained everything in class, and you had one or more mathematics textbooks but you only really used them for the problems, so that you rarely sat down and tried to read the expository sections. If this is your situation, it's probably important to recognize your lack of experience and to acknowledge that learning new mathematics via independent reading might involve developing some new skills.

But, you might ask, why would you need to learn by reading independently? Surely your college professors, like high school teachers, will explain everything to you? Well, yes and no. Yes in the sense that they give lectures, and that professors and other people are available to help you when you get stuck (see Chapter 10). But no in the sense that your lecture classes will sometimes be much larger than your high-school classes, and you will sometimes find that the professor goes too fast for you to understand everything at the time (see Chapters 9 and 13). This means that you will often have to read your lecture notes carefully after the lecture, and learn the mathematics that way.

I myself came a bit unstuck in this respect in my first undergraduate year. From my high-school experience, I was accustomed to learning by doing exercises and solving problems. My (excellent) teacher would give me problems to do, and I would go away and try them, and sometimes he would give me hints when I got stuck. This meant that I learned lots of things by more or less inventing them for myself—I worked out how to apply ideas, how to combine them, how to use diagrams to help me think about things, and all sorts of useful stuff. I thus had lots of problem-solving practice. What I didn't have was practice at learning mathematics by reading about it. Of course, I could have chosen to read my textbook at any point, but this is something I rarely did.

As a result, at the end of my first undergraduate year, I found myself struggling to understand a number of courses, and my Linear Algebra course in particular. I could apply a lot of the procedures for that course, but I didn't really understand what the procedures achieved. Now, I can't recall what prompted me to do this, but eventually I tried reading the relevant bits of my textbook, which for most of the year had sat on my shelf unopened. To my surprise and delight, it contained some good explanations that helped me to get a proper, satisfying understanding of what was going on. Naturally, once I got over the surprise and delight, I just felt embarrassed—of course the book contained decent explanations, that's why textbooks are written.

My point is that you should not expect that all the information you need will come from a person physically talking to you. A lot of it you can learn from things that are written down. In earlier chapters I discussed various tactics you can use for making sense of written mathematics in the familiar form of calculational procedures, as well as in the less familiar forms of definitions, theorems, and proofs. In those chapters, however, we looked at the details. Here I will discuss reading on a larger scale.

7.2 Reading your lecture notes

You shouldn't have any problem acquiring a full set of lecture notes for each of your courses (though see Chapters 9 and 11). Probably, however, you will not understand everything that is in these notes; you will come out of many lectures with only a partial understanding of the new ideas. One thing you should therefore do as you go along is read your notes carefully, think about them hard, and generally try to understand them. I am aware that saying that sounds ridiculous, but you would be surprised at how many students don't do it. I know this because, in my university's Mathematics Learning Support Centre, I regularly see students attempting problems without any apparent prior effort to understand their notes. I have conversations with them that go like this:

STUDENT: I can't do this problem.

ME: Hm, I haven't studied this for a while. What does that word mean?

STUDENT: Um, I don't know.

ME: Well, have you got your lecture notes with you so we can look it up?

STUDENT: Um, no.

It sounds daft, right? The person is trying to answer a question they don't understand, they know they don't understand it, they haven't studied the materials that have been provided to help them understand it, and they don't even have those materials to hand. It might be that they're a genius who can usually reinvent a whole bunch of mathematical theory on their own, but in such cases this doesn't seem likely.

I can, though, see why this happens. I think it is because students are accustomed to learning everything from a teacher's live explanation, so they are not used to reading notes in a systematic way or to treating them as the main source of information. Your professors will expect you to do this reading, and people will give advice like "You can't read a textbook like a novel!" or "You should read with a pencil and paper to hand!" or "Mathematics is not a spectator sport!" In my view, these are sensible suggestions, but they are not specific enough to act as guidelines for what you should actually do when trying to learn from lecture notes or a book. So I'm going to give a lot more detail, demonstrating how I would read an unfamiliar piece of mathematics. Stay with me for a moment, even if that doesn't sound like something you need. I once went through this with a bright and successful mathematics student and, as I did so, his eyes got wider and wider—you could practically see him thinking "Wow, yeah, I can see that I should be doing that, but I guess I never do." I've since done similar things with other successful students, and it does seem that a lot of them don't actually read very well. So read enough of the next section to decide whether your mathematical reading could be improved.

7.3 Reading for understanding

We'll look at the extract below, which comes from a set of lecture notes for a course called Differential Equations. It does not matter if you haven't studied differential equations, because I will explain how I would make sense of this if I were in that position. Skim-read it first, then read the detailed explanation below.

2.2 HOMOGENEOUS EQUATIONS

Differential equations of the form $\dfrac{dy}{dx} = f\left(\dfrac{y}{x}\right)$ are called *homogeneous equations*.

Specifically, this applies to equations that can be written in the form

$$\frac{dy}{dx} = \frac{P(x,y)}{Q(x,y)}$$

where P and Q are homogeneous expressions in x and y and are of the same degree.

For example $\quad \dfrac{P}{Q} = \dfrac{x^3 + 2xy^2 + y^3}{x^2y + yx^2 + 2y^3} \quad$ or $\quad \dfrac{P}{Q} = \dfrac{ax^2 + bxy + cy^2}{lx^2 + mxy + ny^2}.$

Equations of this type can be transformed into separable equations by the substitution $y = vx$ where v is a function of x. With this substitution

$$\frac{dy}{dx} = v + x\frac{dv}{dx},$$

so that the equation becomes

$$v + x\frac{dv}{dx} = f(v)$$

which can be separated as

$$\int \frac{dv}{f(v) - v} = \int \frac{dx}{x}.$$

Example 2.5: Consider the equation $2xy\frac{dy}{dx} = x^2 + y^2$.

[The notes have a box after this, in which the student is supposed to fill in some text, perhaps in a lecture or perhaps on their own.]

I will tell you what went through my head as I tried to read this (I'm not faking—I wasn't into differential equations when I was an undergraduate so I've forgotten most of this stuff).

The first line says

Differential equations of the form $\frac{dy}{dx} = f\left(\frac{y}{x}\right)$ are called *homogeneous equations*.

The first thing I notice is that I'm looking at a definition. It isn't labeled as such, but the phrasing and the italics give this away. The new thing being defined is a new type of differential equation, and this definition indeed involves a derivative dy/dx being equated to some function $f(y/x)$. Normally we just see functions written as $f(x)$, so it looks like the y/x part must be what makes this a special kind of differential equation. One other thing I notice at this point is that the derivative is dy/dx, which means that we must be able to think of y itself as a function of x. That is what we'd expect, and a solution to the differential equation will be a specific way of expressing y as a function of x (see Section 2.6).

The second line says

Specifically, this applies to equations that can be written in the form

$$\frac{dy}{dx} = \frac{P(x, y)}{Q(x, y)}$$

where P and Q are homogeneous expressions in x and y and are of the same degree.

The "specifically" here makes me think that this must be a special case of the general definition, and I can check this: we have $\mathrm{d}y/\mathrm{d}x$ on the left-hand side, and some function of x and y on the right, which is what I would expect (presumably a differential equation can be homogeneous in other ways too but we are not considering those at this stage). The function on the right-hand side is written in a very general form as a quotient of two functions of x and y, so it is not immediately clear to me how it relates to the general form $f(y/x)$. Also, I don't know what it means for P and Q to be homogeneous expressions in x and y that are of the same degree. But there seems to be an example coming up, so perhaps both of these things will become apparent if I carry on.

The third line says

For example $\quad \dfrac{P}{Q} = \dfrac{x^3 + 2xy^2 + y^3}{x^2y + yx^2 + 2y^3} \quad$ or $\quad \dfrac{P}{Q} = \dfrac{ax^2 + bxy + cy^2}{lx^2 + mxy + ny^2}.$

Clearly these are intended to be examples of the general expression in the second line, and they do clear up the second of my two uncertainties: P and Q in both cases are expressions in x and y, and all the terms have the same combined degree (three in the first case, and two in the second). So the degree part makes sense, but I'm still not sure what makes $x^3 + 2xy^2 + y^3$, for instance, a homogeneous expression. At this point I could look this up, or continue. I think I will continue, because it's pretty clear what type of function I'm supposed to be considering. But I would make a note to look it up or ask later.

What's not clear is how P/Q in either case is a function of y/x. They both seem to be functions of y and x, but that is not quite the same thing. Perhaps it is always possible to rearrange a function like this to write it as a function of y/x? I can't immediately see how to do this with either of the examples given, although it seems plausible that the special structure of P/Q in both cases might mean that something like this is always possible. Perhaps it would be easier to start by building up simple functions of y/x and seeing what I can get. Trying this, I can make the following functions, by squaring, by taking a reciprocal, by adding the results of these together, and by taking the reciprocal of the result:

$$\frac{y^2}{x^2} \qquad \frac{x}{y} \qquad \frac{y^2}{x^2} + \frac{x}{y} = \frac{y^3 + x^3}{x^2 y} \qquad \frac{x^2 y}{y^3 + x^3}.$$

Looking at these, it certainly seems plausible that I could make functions P/Q like those given in the example, so I'm satisfied enough at this stage to move on.

The fourth (long) line says

Equations of this type can be transformed into separable equations by the substitution $y = vx$ where v is a function of x. With this substitution

$$\frac{dy}{dx} = v + x\frac{dv}{dx},$$

so that the equation becomes

$$v + x\frac{dv}{dx} = f(v)$$

which can be separated as

$$\int \frac{dv}{f(v) - v} = \int \frac{dx}{x}.$$

In this part we seem to have moved away from the examples and back to a general discussion. It looks like it might be fairly straightforward to understand, so I will work on it in isolation first, before trying to relate it to the examples. Looking at $y = vx$, I can see that to find dy/dx we are differentiating a product, so using the product rule we will get v times x differentiated, plus x times v differentiated. That does indeed come out as stated, because v is a function of x. Notice that this is important—if v were just a constant, it would differentiate to 0, so we need to keep track of what is being treated as a function of what. Then I can look at the next equation and see that this simply involves replacing dy/dx with $f(v)$. I was expecting it to be replaced with $f(y/x)$, but with a moment's thought I can see that the substitution $y = vx$ does give us $v = y/x$. To get to the last equation I need to do a bit of algebra, which I'd probably write out to convince myself that it seems okay:[1]

$$v + x\frac{\mathrm{d}v}{\mathrm{d}x} = f(v) \Rightarrow x\frac{\mathrm{d}v}{\mathrm{d}x} = f(v) - v$$

$$\Rightarrow \frac{1}{x}\frac{\mathrm{d}x}{\mathrm{d}v} = \frac{1}{f(v) - v} \Rightarrow \int \frac{\mathrm{d}v}{f(v) - v} = \int \frac{\mathrm{d}x}{x}.$$

What I don't know at this point is how any of this would apply to particular functions like those listed as examples above. I'm not sure I really want to think about that, either, since those functions looked rather complicated. I think I'll have a better chance of working it out in relation to the considerably simpler example at the end of the extract, so I'll try that instead:

Example 2.5: Consider the equation $2xy\dfrac{\mathrm{d}y}{\mathrm{d}x} = x^2 + y^2$.

I can see that this equation is not quite in the form I've been looking at, but by rearranging it in the obvious way I can put it into that form:

$$\frac{\mathrm{d}y}{\mathrm{d}x} = \frac{x^2 + y^2}{2xy}.$$

Now I'd like to work out how I can apply the general solution process to this particular differential equation. I'm not sure what role v is playing, but I can temporarily suspend that concern and just go ahead and make the substitution $y = vx$. This means that I can replace $\mathrm{d}y/\mathrm{d}x$ on the left-hand side as in the general case, and also replace y by vx on the right-hand side, to give

$$v + x\frac{\mathrm{d}v}{\mathrm{d}x} = \frac{x^2 + v^2x^2}{2vx^2}.$$

Now I can suddenly see the point of this substitution: all the instances of x^2 on the right-hand side cancel, leaving a much simplified equation:

$$v + x\frac{\mathrm{d}v}{\mathrm{d}x} = \frac{1 + v^2}{2v}.$$

---

[1] In fact, because of my extra training, I'm aware that it is sometimes dodgy to treat a derivative as a fraction that can be manipulated in this way but, since this comes from a set of lecture notes, it seems safe to assume that it must work in this case.

I can also now see why we need P and Q to have the stated form: when they do, making this substitution will always mean we end up with the same power of x everywhere, so such canceling will always work. Thus, already, this example has provided me with much improved insight into why the general formulations are as stated. Then I can go ahead and do the algebra for this particular case, first rearranging to give:

$$v + x\frac{\mathrm{d}v}{\mathrm{d}x} = \frac{1+v^2}{2v} \quad \Rightarrow \quad x\frac{\mathrm{d}v}{\mathrm{d}x} = \frac{1+v^2}{2v} - v = \frac{1+v^2-2v^2}{2v} = \frac{1-v^2}{2v};$$

then separating, to give:

$$\int \frac{2v}{1-v^2}\,\mathrm{d}v = \int \frac{\mathrm{d}x}{x}.$$

At this point I know that I would be able to find both integrals, which would give an equation for v in terms of x, which I could then convert to find y in terms of x. You might like to pause here and do that. I will go on to discuss what I'd like you to observe about this explanation.

The first thing is that reading lecture notes is not necessarily easy. Notes often start with rather general formulations, and they do not necessarily give every bit of explanation you might need in order to see how everything is linked together. This, of course, is why you should go to lectures. Doubtless many of the questions I raised for myself would have been answered by the professor in his verbal explanation.

The second thing to observe is that my reading process has a number of characteristics. First, it involves identifying the status of the thing I'm looking at: is this a definition, theorem, example, proof, or something more general like a motivating introductory paragraph? Knowing the status of a particular item means I know what kind of information to expect from it. Second, when I come across a word for which I can't remember the meaning, I'll look up the appropriate definition or make a note to do so later. Third, when I come across an example, I'll check that the definition or property does seem to apply. (Similarly, if I come across a reference to an earlier theorem—"by Rolle's Theorem" or "from Theorem 3.1" or "using Lemma 2.7"—I'll look that up to make sure I know what it says and can see how it applies in the current situation.) Fourth, I'll try to understand a general statement both in the abstract and in relation to some examples (see also Chapter 2). Fifth, if examples are hard to think of or if the provided ones look complicated, I'll look ahead, because perhaps the professor has provided a simple example somewhere after the general statement. This happens a lot. Mathematicians know that general statements can be hard to understand, so they often give a straightforward example fairly soon afterwards.

An overall characteristic of this reading is that some of it involves looking back and some of it involves looking ahead; I jump around much more than I would do if I were reading a different kind of text. I'd like to be able to describe how I know when to do what—when I consider it reasonable to think "Well, never mind, I'll move on now and see if that makes sense a few lines further down." But I don't really know how I make those decisions, I'm afraid—I think I do it more on the basis of a feeling about what I need rather than a rule that tells me how to behave. Nonetheless, I hope that seeing an illustration of my reading process might help you see what you could do to make sense of lecture notes. And this seems like a good time for a reminder of a piece of advice from Chapter 3: you must read the words. You really must. Think how much information you'd have missed from that extract if you'd just skipped ahead to find the nearest calculation.

7.4 Reading for synthesis

When I say reading for synthesis, I mean reading to understand how the mathematics covered in a whole course fits together. This is desirable pragmatically because it will help you to digest the material in preparation for an exam. It's also desirable intellectually, because one of the joys of mathematics lies in seeing the relationships within a whole area of mathematical theory.

To some extent, getting a sense of how a course fits together will be natural—you will notice that there are similarities between various calculations, that some theorems are proved using earlier theorems, and so on. Indeed, if you are listening properly (see Chapter 9) you will find that your professor gives a lot of information to help you recognize this kind of link. Nonetheless, you might be able to improve your overall grasp of what's going on by making some overt effort in this direction. I have some specific suggestions regarding how you might do this; if you follow these suggestions as a course progresses, you will find that you've already built up a good set of revision notes by the time you want to start your exam preparation.

The first strategy is to keep a running summary of what is in the lecture notes. You might be provided with something like this at the beginning of a course, in the form of a contents page or a course outline (this is getting more common now that so many professors provide some form of printed notes). However, a contents page probably isn't quite what you want, because it will include chapter headings and subheadings but not the actual content of definitions and theorems, and so on. I mean something that might look more like this (I'm imagining this is the beginning of a section of notes that is about relations):

Definition: relation (notation for general relation is $\sim$)

Examples of relations ($=$, $<$, etc.)

Definition: symmetric relation ($a \sim b \Rightarrow b \sim a$)

Examples ($=$ symmetric, $<$ not symmetric)

Definition: reflexive relation ($a \sim a$)

Examples ($=$ reflexive, $<$ not reflexive)

Definition: transitive relation ($a \sim b$ and $b \sim c \Rightarrow a \sim c$)

Examples (both $=$ and $<$ transitive)

I'm abbreviating here in a way that I never would in a piece of mathematical writing for someone else (see Chapter 8 for a discussion of formal writing). I'm summarizing for myself, so I'm including just enough information to remind myself what each bit of mathematics is about. I've got the status of the item (is it a definition, a theorem, a set of examples, etc.?) and a brief note to remind me of at least some of the main ideas. If you're feeling super-keen, you might like to supplement this sort of summary with page numbers so that you can easily move between the summary and your notes.

As well as an overall summary, you might also like to keep a separate definitions list. This can save you a lot of time when doing problem sets and the like. For the reasons discussed in Chapter 3, you will need to use definitions frequently and, if you keep having to search through your notes to find them, you'll waste a lot of time. You do need to make sure your definitions are absolutely logically correct, so in this case I would write each one out in full (the aim is different from that of the summary list: accuracy instead of brevity). Keep the definitions list at the front of your notes and then you can refer to it easily. You might also keep a theorems list, if that looks like being useful. Personally I would keep this separate from the definitions list, because of the different status of definitions and theorems within mathematical theory. But you might find that different approaches work in different courses.

Making abbreviated lists should help you to get an overview of what's in a course without getting bogged down in the detail. This might be enough for you, especially if you are a list-maker anyway. Personally, I'm okay with lists but what I really like is a good diagram. As a result, I tend to prefer *concept maps* (you might have heard these called *spider diagrams* or *mind maps*). Here, for instance, is part of a concept map I made for my recent course in Multivariable Calculus (you might not know what all the writing in the boxes means, but it should give you an idea about how I lay such things out):

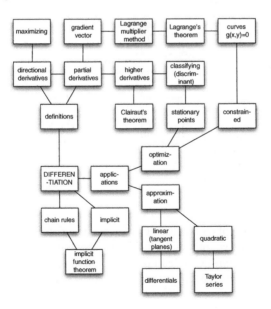

I drew this in front of the students during a revision lecture. I drew the branches bit by bit, and my hope is that this made clearer to the students that, although the course was "large," it could be thought of in terms of key topics, and in terms of different procedures that achieve similar aims.

Multivariable Calculus (usually) involves some theorems and proofs and numerous procedures for calculating derivatives and integrals. It is a bit of a hybrid in that sense. Concept maps can be even more useful for mapping out a fully theoretical proof-based pure mathematics course, because then one can use arrows to capture logical dependence; to indicate which definitions and theorems are used to prove which other theorems. Here, for example, is part of a concept map for the continuity topic in a course called Analysis. In this one, I also used different shapes for the definitions (ovals), examples (octagons), and theorems (rectangles):

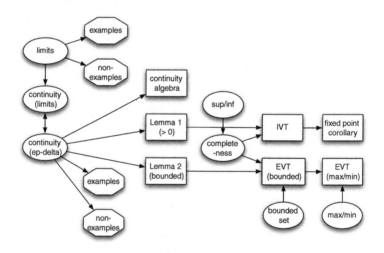

I gave out copies of the whole map at the beginning of the course so that students could keep track of the build-up of the course. As far as I know, not many professors do this sort of thing. In fact, I might have done my students a disservice, because maybe it is better for a student to produce a concept map for themselves. Certainly making a good concept map is harder than making a good list. In the case of a list, you'll just be recording what's in the course in the order in which it appears, and your professor decides that. In the case of a concept map, you'll have to look for what is linked to what, and probably make some decisions about which of those links are worth recording. That's exactly the sort of thing your professors would want you to do.

Finally, while I'm talking about reading for synthesis, I'd like to say a word about highlighting and color-coding. A lot of students seem to be somewhat addicted to highlighters. Their notes have highlighting all over every page, sometimes in a variety of colors. They seem to use highlighting not to mean "this is important" so much as "I have seen this line!" I don't suppose there's anything wrong with that—I personally don't like it because I think it makes the pages look garish and hard to read, but I can see that some people might think it attractive. But I would be concerned if I thought that a student had missed an opportunity to use the highlighting to mean something. Perhaps, for instance, to highlight all the definitions in pink and all the theorems in green. Or to organize, not by status, but by concept: differentiation in orange and integration in blue, perhaps. It seems to me that such a strategy might facilitate memory in some useful way. The same goes for writing different bits of notes in different colored ink; if you're going to use color, it's probably worth having a system.

7.5 Using summaries for revision

I used to find lists and concept maps very useful at revision time. During my first undergraduate year, I developed a habit of condensing each course into a list on one side of paper. Sometimes I had to condense to two sides first and then condense further; doing so meant abbreviating a lot, while retaining enough information that I would know what everything meant. But it wasn't that hard—I didn't have to read the eventual list with a magnifying glass. I didn't start using concept maps until a bit later, but by the time I did my masters degree I was specializing in pure mathematics, and at the end of every course I would make a concept map and stick it up on the fridge or a kitchen cupboard door or somewhere where I would see it all the time. I'm not sure whether it was making the map in the first place that helped, or seeing it all the time, but the whole process certainly gave me a better sense that I was familiar with the structure of the whole course. These days I usually advise students not only to make some kind of summary, but also to use it in a particular way. Here is that advice.

When you want to start revising, go through your summary and, next to each item, write either a check (✓) to indicate that you are fine with it, or a question mark to indicate that you sort of get it but you're not really sure, or a cross to indicate that you haven't a clue what it's about. Doing this might be a bit alarming, because you'll start out with a fair number of question marks and crosses. But it's better than deluding yourself about what you do and don't know, because it puts you in a position to work effectively to improve the situation. Then ask yourself: which items should you work on? Quite a lot of students say "those with crosses." That's a sensible intuitive answer, but I do not think it is likely to be maximally effective. I think you should work on the items with question marks, for two reasons. First, you already know a bit about these, so you are likely to make progress. This means that you will add things to the checks pile reasonably fast, which is a good idea because (as discussed further in Chapter 13) it might be better to know some things well rather than knowing everything badly. It'll also make you feel good, which will keep you more motivated. Second, as you work on the items with question marks, some of the things with crosses will become question marks without you having to do anything at all. This happens because you are increasing your knowledge base, so some of the things that previously seemed meaningless start to seem accessible after all. Of course, at some point you might still get a bit depressed about all the things you don't know yet. If that happens, you can always cheer yourself up by working on your checked items for a while.

Now, these are strategies that worked for me, and I'm not claiming that they will necessarily work for you. I don't know how effective they are on average. It might be that you already have strategies that are more effective than these although, in that case, you should think about whether they will transfer well to upper-level mathematics. And I do think it's probably good advice to listen to things that have worked for people who've been successful, so ask your professors or tutors what they think too.

7.6 Reading for memory

No two ways about it: when studying for a major you are going to have to remember a lot. One glorious thing about mathematics, of course, is that much of the time this will be effortless. When you understand a piece of mathematics well, you can often just remember it immediately because it seems like everything naturally has to be that way. Even for complicated things, I find that I can often reconstruct what I need by deriving it from more familiar information. But you will have to take exams, which means that you will, at some point, have to have a lot of mathematics at your fingertips. In this section I'm going to discuss some ways of making this happen.

One thing to do, which works particularly well for procedures, is to practice until whatever you're doing becomes routine. This will sometimes work well in advanced mathematics, as it no doubt did for you in high school. However, as discussed in Chapter 1, the procedures you'll be learning will be longer, they'll involve more decisions, and you'll have fewer examples to mimic and fewer set exercises. So you shouldn't expect that it'll be quite as easy to get the practice as it might have been earlier in your education. You might have to go a bit further out of your way to find practice problems, using more books, and so on. Or, also as discussed in Chapter 1, you might consider constructing exercises for yourself.

Another way to remember things is to relate them to other things. Sometimes, in this respect, mathematical language does us a favor. For instance, recall that in Chapter 3 we looked at a definition of *increasing* for functions. You could, if you wanted to, memorize that definition by rote. You could write it on a card and test yourself lots of times until you

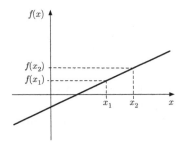

got it right. But I think that would be pretty inefficient. *Increasing* means more or less what most people think it ought to mean, so I wouldn't try to memorize it, I'd try to reconstruct it by capturing my intuitive understanding instead. Probably, because I like diagrams, I'd remember a diagram like the one on the previous page, and reconstruct it from that.

Have a go now. Can you do it? If so, great (but perhaps do check that you did write something logically equivalent to my version, just in case). If not, look back at the definition in Section 3.6, and make sure you can see how each part relates to the diagram. Then try again tomorrow.

Similar advice applies for theorems. Sometimes theorems state rather complicated things that you would not have thought of for yourself. In those cases, a bit of deliberate memorization might be in order, at least until you have a reasonable grip on a course. But sometimes theorems state simple things that are quite intuitively obvious. In those cases, it shouldn't be too hard to reconstruct the theorem. Consider, for instance, the Intermediate Value Theorem (you will probably meet this in a course called Analysis). This is often accompanied by a diagram like the one below, and it says:

Theorem: Suppose that f is continuous on $[a, b]$ and that y_0 is between $f(a)$ and $f(b)$. Then there exists $x_0 \in (a, b)$ such that $f(x_0) = y_0$.

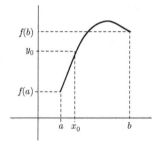

First, try applying the skills discussed in Chapter 4 to this theorem. Can you see how everything in the diagram relates to it? Does the theorem seem, to you, to be true? Can you imagine varying the function so you can check that the conclusion always seems to hold? What if we relaxed the premise that f has to be continuous—does the conclusion necessarily follow? Can you see why it is called the Intermediate Value Theorem? Then try covering up the theorem and reconstructing it while looking at the picture.

Here is some advice about doing this kind of theorem reconstruction. Often it seems to be the case that remembering the conclusion is easier. You might, for instance, look at the diagram above and first think to write $f(x_0) = y_0$. There is no reason why you can't write that down first, leaving a line or two of space so that you can then arrange the rest of the theorem around it. Then, we need the premise, and (as in Chapter 4) a premise should introduce the objects we're working with. Here we need to specify where y_0 is, which would lead us to write something like this:

[gap] Suppose that y_0 is between $f(a)$ and $f(b)$.

[gap] $f(x_0) = y_0$.

Then we can note that x_0 needs to be introduced, and decide what to say about it. Clearly x_0 is on the x-axis, we can't tell exactly where it's going to be except that it's between a and b, and clearly it's not the case that every x_0 has $f(x_0) = y_0$. So it must be sensible to say there exists an x_0 with this property. That finishes off the conclusion:

[gap] Suppose that y_0 is between $f(a)$ and $f(b)$.

Then there exists $x_0 \in (a, b)$ such that $f(x_0) = y_0$.

It doesn't quite finish off the premises because we need to introduce the function and its properties. If we've thought properly about the theorem in the first place, it will probably come to mind that the function needs to be continuous—otherwise there might be a jump in the function graph and the conclusion would not hold. So now we can complete the theorem:

Suppose that f is continuous on $[a, b]$ and suppose that y_0 is between $f(a)$ and $f(b)$.

Then there exists $x_0 \in (a, b)$ such that $f(x_0) = y_0$.

In fact, if you use other cues then reconstructing a theorem might be even easier. In Analysis, where you will probably meet this theorem, some main function properties studied are continuity, differentiability, and integrability. You might be able to work out that we only need continuity in this case. But, if you've done your summarizing well, you might not even need to do that; you might just remember that this theorem appeared at the end of the continuity chapter, so that must be all we need. Some students might think that using such memory tricks is somehow cheating; that a dedicated student would just learn it all properly to begin with. I suppose that might be true. But I don't think it is, because successful mathematicians are smart and strategic. They don't want to spend time memorizing things when they could reconstruct them using a bit of effort and common sense.

There is one thing I do want to emphasize, though, about remembering things by reconstructing them: you have to give yourself a bit of time. I say this because sometimes, when I ask a student a question, I see them reaching for their notes before they've even thought about it. This makes me sad. Sometimes I stop students as they're doing this, and ask if they can pull the mathematics out of their memory instead. I would say that more than half of the time, they can. It might take them a minute (I mean literally a minute) and they might have to remember just part of it first and then reconstruct the rest. But a minute is no longer than it would have taken to look it up and, more to the point, succeeding tends to improve their confidence.

7.7 Using diagrams for memory

You might have noticed that in these sections about memory, I've been talking a fair bit about diagrams. We have to be careful with diagrams, because it is easy to draw something that is not quite general enough and then to be misled by it, and because a diagram does not replace a formal proof. But diagrams can capture a lot of mathematical relationships in a memorable format.

For a simple example, consider some standard relationships involving sines, cosines, and tangents of angles. I don't know about you, but I can never remember which angle has sine equal to $\sqrt{3}/2$ and which has sine equal to $1/\sqrt{2}$. Fortunately, I don't need to. I can work it out by drawing the triangles in the diagram below and using Pythagoras' Theorem to work out the lengths of the remaining sides. Sometimes it takes me a couple of attempts to decide which length should be 1, but I still find this less effort than remembering all the sines, and so on, separately; I don't like remembering "facts"—I find it difficult, tedious, and unrewarding.

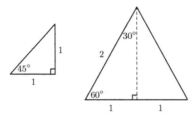

Even better, in my view, is the diagram below, showing a unit circle, an angle θ measured clockwise from the positive x-axis, and a right-angled triangle.

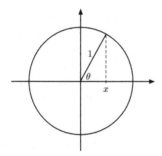

What can we see on this diagram? Many, many things. First, the point marked x is equal to $\cos \theta$. Can you see why? Think about the side adjacent to θ in the right-angled triangle, for which the hypotenuse is of length 1. We can see $\sin \theta$ on the y-axis in a similar way. Second, we can see that $\cos 0 = 1$ and $\sin 0 = 0$, which can be handy, especially for those who are prone to forgetting where the sine and cosine curves cross the axes. Third, it can help in remembering which of $\cos \frac{\pi}{6}$ and $\sin \frac{\pi}{6}$ is equal to 1/2. For the finale (although no-one ever seems to forget this one, to be fair) the diagram makes it obvious that we always have $\cos^2 \theta + \sin^2 \theta = 1$. This is one of my favorite diagrams and I'm always sad to meet undergraduate students who are not familiar with it, because it saves me so much work.

7.8 Reading proofs for memory

In this final section I will talk about remembering proofs. Students are often concerned about this aspect of their studies. It seems to them that there are many proofs, that they are all different, and that quite a few of them are long. This starts, then, to seem like a difficult memory task, especially in courses in which students are expected to produce standard proofs in an exam. I'm not going to insult you by pretending that it isn't difficult. But you can probably make it a lot easier for yourself by adapting some strategies that are familiar from your work with mathematical procedures.

First, notice that when you remember a procedure, you don't remember every detail of a particular use of it, you just remember the main steps. For instance, if you had to tell someone how to locate the local maxima and minima of a function, you'd probably say something like "differentiate it, then set the derivative to zero and find the solutions, then find the second derivative for each solution; if it's positive, the point is a minimum, and vice versa." You wouldn't say anything about how to differentiate, or about the derivatives of particular functions; you would just describe the major steps and treat the rest as routine calculations.

This can also be done with less obviously procedural things. For instance, if you've done some mechanics, you've probably solved problems involving particles at rest (on a slope with friction, that kind of thing). Your set of steps for this kind of problem is probably something like "draw a diagram to represent the physical situation, mark all the forces on it with arrows, decide which directions to work with (horizontal/vertical or down the slope and perpendicular to that), then equate forces in those directions." Again, how this plays out in any particular problem is a matter of detail. If you have an overview of the process, you can work out that detail for any particular case as you go along.

Constructing and reconstructing proofs can be a bit like this. Getting started can be quite mechanical. When proving that a definition is satisfied, we can often use that definition to structure a proof, and when proving that a more general theorem is true, the structure of that theorem can help us in a similar way (see Chapter 5). We can always try writing premises and conclusions in terms of definitions, and there are several standard proofs types and tricks that are useful for proving particular types of statement (see Chapter 6). In all, we can treat proving as a sort of higher-level procedure: "ascertain the structure of what is to be proved, write premises and conclusions in terms of definitions, consider standard proof types (if get stuck, try thinking about specific examples)."

If you think of proving in this way, and if you are on the lookout for places in which standard structures and tricks are used, you should find that even long proofs don't seem too mysterious. At the undergraduate level, it's a rare proof that involves more than one good trick. So you can probably reduce most of the proofs you see to a compressed version along the lines of "write in terms of definitions, perform trick of adding and subtracting $f(x + h)g(x)$, split up pieces, and manipulate algebra in obvious way" (that was a description of the proof of the product rule for differentiation—did you recognize it?). If you can do this for most of the proofs you encounter, you will find that even a proof-heavy course doesn't have that many ideas in it. Most of the 10- or 15-line proofs you see can probably be reduced to two or three main steps, so that you only need to remember these main steps in order to reconstruct the whole thing. You'll have to fill in some detail in terms of algebraic manipulations or other calculations, but you're used to doing that with procedures anyway. So being able to reconstruct proofs in exams is not easy but, if you reduce the workload by thinking in terms of the main steps, it needn't be impossible either.

As a final point in this chapter, I have one more thing to say about reading for memory, especially when it comes to fairly long proofs: be very, very careful if you decide you're going to memorize something by rote. I hope you won't do that anyway, because it's not a rewarding use of time. But if you do try, you should be aware that for a professor, it is often very obvious when someone has done this. Sometimes, on a student's exam, I see a proof that looks alright when I hold it out at arm's length, but that makes no sense at all when I actually start to read it.

Sometimes the logic is all over the place, and key ideas are missing or are presented in strange, illogical orders. Sometimes notation is used before it is introduced, or introduced then never used, or introduced to mean one thing but apparently used to mean another. I feel bad when I see this kind of thing, because clearly the student has made a genuine effort to remember this piece of mathematics. But, by writing it so badly, they have demonstrated that they do not understand it. I cannot, then, give them a good grade, because understanding is what I value. So please try to exhaust all approaches that might lead to understanding before you resort to rote memorization—there are many things you could do that would be much more satisfying.

SUMMARY

- Studying as a mathematics major will involve independent reading; your experience of this might be limited so it might involve developing new skills.
- You will probably not understand everything in your lectures, so you will have to study your lecture notes; students who do not do this often find it difficult to get started on problems.
- Reading lecture notes is not necessarily easy; notes usually contain a combination of general explanations and examples, and it might take some work to establish the links between them.
- Reading mathematics should involve identifying the status of different items, looking up earlier definitions and theorems, relating examples to general statements, and looking back and ahead more than is usual in ordinary reading.
- Various types of summary can be useful when reading for synthesis: you might want to consider keeping contents lists, definitions lists, and concept maps.
- Summaries can be used to aid revision planning; I suggest a check/question-mark/cross system, and working first on the question-mark items.
- Practice has its place in reading for memory, but you should aim for a situation in which you can remember a diagram or the key ideas of a definition, theorem, or proof, and reconstruct the rest.

FURTHER READING

For more on reading mathematical proofs, try:

- Solow, D. (2005). *How to Read and Do Proofs.* Hoboken, NJ: John Wiley.
- Allenby, R.B.J.T. (1997). *Numbers & Proofs.* Oxford: Butterworth Heinemann.

For more on preparation for tests and exams, try:

- Moore, S. & Murphy, S. (2005). *How to be a Student: 100 Great Ideas and Practical Habits for Students Everywhere.* Maidenhead: Open University Press.
- Newport, C. (2007). *How to Become a Straight-A Student: The Unconventional Strategies Real College Students Use to Score High While Studying Less.* New York: Three Rivers Press.

CHAPTER 8

Writing Mathematics

This chapter is about the importance of good mathematical writing, and why you should aim not only for correct answers but also for professional presentation of your ideas. It explains how to make sure your mathematical arguments are presented clearly, and gives tips on avoiding common mistakes when using new mathematical symbols.

8.1 Recognizing good writing

Consider the following two calculations, the author of each of which is trying to find

$$\int \sin^4 x \, dx.$$

Do not focus on whether you know a different (perhaps better) way of doing the calculation. Instead, think of yourself as a reader, and pay attention to how much work you have to do to follow each person's thinking.

CALCULATION 1

$$\int \sin^4 x \, dx \qquad\qquad u = \sin^3 x \quad \frac{dv}{dx} = \sin x$$

$$\frac{du}{dx} = 3 \sin^2 x \cos x \quad v = -\cos x.$$

$$= -\sin^3 x \cos x - \int -3 \sin^2 x \cos^2 x \, dx$$

$$= -\sin^3 x \cos x + 3 \int \sin^2 x \cos^2 x \, dx$$

$$= -\sin^3 x \cos x + 3 \int \sin^2 x (1 - \sin^2 x) \, dx$$

$$= -\sin^3 x \cos x + 3 \int \sin^2 x \, dx - 3 \int \sin^4 x \, dx$$

$$\cos(2x) = 1 - 2 \sin^2 x$$

$$\sin^2 x = \tfrac{1}{2}(1 - \cos(2x))$$

$$\int \sin^2 x \, dx = \tfrac{1}{2} \int 1 - \cos(2x) \, dx = \tfrac{1}{2}x - \tfrac{1}{4}\sin(2x)$$

$$-\sin^3 x \cos x + 3 \int \sin^2 x \, dx - 3 \int \sin^4 x \, dx$$

$$= -\sin^3 x \cos x + 3 \left(\tfrac{1}{2}x - \tfrac{1}{4}\sin(2x)\right) - 3 \int \sin^4 x \, dx$$

$$= -\sin^3 x \cos x + \tfrac{3}{2}x - \tfrac{3}{2}\sin x \cos x - 3 \int \sin^4 x \, dx$$

$$4 \int \sin^4 x \, dx = -\sin^3 x \cos x + \tfrac{3}{2}x - \tfrac{3}{2}\sin x \cos x$$

$$\boxed{\tfrac{3}{8}x - \tfrac{1}{4}\sin^3 x \cos x - \tfrac{3}{8}\sin x \cos x + c}$$

CALCULATION 2

Let $I = \int \sin^4 x\, dx$.

Let $u = \sin^3 x$ so $\dfrac{du}{dx} = 3\sin^2 x \cos x$ and $\dfrac{dv}{dx} = \sin x$ so $v = -\cos x$.

Then integrating by parts gives

$$I = -\sin^3 x \cos x - \int -3\sin^2 x \cos^2 x\, dx$$

$$= -\sin^3 x \cos x + 3\int \sin^2 x \cos^2 x\, dx$$

$$= -\sin^3 x \cos x + 3\int \sin^2 x (1 - \sin^2 x)\, dx$$

$$= -\sin^3 x \cos x + 3\int \sin^2 x\, dx - 3\int \sin^4 x\, dx$$

$$= -\sin^3 x \cos x + 3\int \sin^2 x\, dx - 3I$$

So $4I = -\sin^3 x \cos x + 3\int \sin^2 x\, dx$ $(*)$

To find $\int \sin^2 x\, dx$, note that

$$\cos(2x) = 1 - 2\sin^2 x \quad \text{so} \quad \sin^2 x = \tfrac{1}{2}(1 - \cos(2x)).$$

So, from $(*)$,

$$4I = -\sin^3 x \cos x + 3\int \sin^2 x\, dx$$

$$= -\sin^3 x \cos x + \tfrac{3}{2}\int 1 - \cos(2x)\, dx$$

$$= -\sin^3 x \cos x + \tfrac{3}{2}x - \tfrac{3}{4}\sin(2x) + c.$$

$$= -\sin^3 x \cos x + \tfrac{3}{2}x - \tfrac{3}{2}\sin x \cos x + c.$$

So $I = \tfrac{3}{8}x - \tfrac{1}{4}\sin^3 x \cos x - \tfrac{3}{8}\sin x \cos x + c.$

As a reader, which calculation would you say is better? To a mathematician, it's obviously the second one. The two arguments contain essentially the same information, but the second one is much easier to read. It has a few words to explain what is going on at each stage, so the reader can see why certain manipulations are valid and/or sensible. It also gives a name to the original integral and uses this throughout, and uses an asterisk to indicate where different equations are being combined. The first calculation isn't wrong, but it jumps around without explanation, so the reader has to do a lot of work to see how it all fits together. A mathematician would believe that it's not the reader's job to do this—that the writer can and should do more to help them out.

Now, which of these looks more like the way you write your own mathematics? If you are a typical new mathematics major, and if you're not an inveterate liar, you might have to say that it's the first one. I don't say this to be discouraging, quite the reverse: I want you to be aware that, even if you don't always write well, you do recognize what good mathematical writing looks like.

Even with this in mind, students are often surprised by the focus on writing in advanced mathematics. It prompts some to go around saying things like "I can't believe we have to do this—I chose math because I didn't want to write essays!" Such people are being over-dramatic, of course. No-one is asking you to write essays. They are just asking you to make your logic clear. I've already talked about this—good presentation can make your thinking clearer to someone else, and can help you to avoid ambiguity. In this chapter, I will review some of those ideas, and describe some common errors so that you can recognize and avoid them.

8.2 Why should a student write well?

If your teachers and professors have already taught you to focus on mathematical writing, then much of the advice in this chapter might be familiar. If not, don't worry—it isn't hard to write well, it's just considered more important in upper level courses than it typically is in earlier education. A word of warning before we start, however: you may think that some of the advice in this chapter is finicky and pedantic. If so, bear with it: you don't want to be considered a poor mathematical thinker when you are actually a good thinker and just a poor writer. To further motivate you to make the effort, I will offer you three reasons for writing well.

The first reason is a short-term pragmatic one: writing well will get you better grades. If you do it, you will do better in coursework assignments and exams. Some students, when they start taking upper-level courses, think this is a bit unfair. They think that, because this is mathematics, the emphasis should be on getting the right answer. To explain why it is not unfair, I usually ask them to imagine that someone has given an *incorrect* answer. This happens quite a lot, especially in exams. Now put yourself in my shoes, as a professor doing the grading. If a student has given an incorrect answer, but has written their argument so that all their logic is clear, then I can say with some confidence that they knew what they were doing, and that they have just made a minor error (lost a minus sign, or something). In that case, I'm often happy to give them most of the points. On the other hand, if a student has given a badly-written incorrect answer, often I cannot tell whether they knew what they were doing or not. This is especially true if they have written bits of their argument all over the place, so that I don't know how it is all supposed to fit together. It might be that they knew what they were doing, but it might be that they had no idea and were just writing down random things in the hope of getting partial credit. Should I give them the benefit of the doubt? If you were the student who had written such an "argument," you might be inclined to say yes. But if you were a student who got the same wrong answer but wrote the argument well, you'd probably think that the person who wrote badly did not deserve the credit—that your argument shows better mathematical understanding. That's the kind of judgment I have to make all the time. If I can't tell whether someone's thinking was reasonable, then I can't in good conscience give them a good grade. It wouldn't be fair to the other students.

The second reason for writing well is also pragmatic but is more long-term, and might appeal to you especially if you intend, one day, to make a lot of money. It is this: whatever career you go into, it will involve communicating with others in writing, and if you do not do this well you will find it difficult to be successful. If you become an accountant, you will have to produce reports in standard language so that different parts of your company can understand the financial situation as it pertains to them. If you become an actuary, you will have to work out how your client companies should invest their employees' pension contributions, and provide recommendations explaining the advantages and risks associated with various options. If you become a logistics manager for a large retail company, you will have to make decisions about how to move your stock around efficiently, and you will have to write documents that explain how your system operates and why it is a good one. If you become an entrepreneur, you will have to produce documents for banks and potential investors that make a convincing case for the utility of your products or services and that detail how you and they can make money by providing them. If you become a hedge fund manager, you will have to produce documentation for your clients to explain and justify your investment decisions. If you become a statistician for the government or the weather service or a major drugs company, you will have to design and test statistical models to predict and explain real-world phenomena, and you will have to justify why your model is better than other models and why your seniors should listen to you if they want to improve their predictions. If you become a teacher, you will have to give daily presentations to students of various ages and abilities, and to communicate in writing with parents, school boards, managers, and so on. You see my point. If you want to be employable and you want to be successful, then obviously it's important to be good at working out how to solve problems. But it's also important to be able to communicate convincingly about whatever you think is the best solution. Intelligent people are not impressed by poorly constructed or poorly presented arguments, mathematical or otherwise. So it's a good idea to get some practice now by learning to write in a way that satisfies your professors.

The third reason for writing well is what I would consider the real one, and it is about intellectual maturity in relation to an academic field. At this stage in your education you are supposed to be developing knowledge, not only of mathematics, but also of the working practices of professional mathematicians. That means understanding their thinking processes, but it also means understanding how they communicate with each other, and learning to communicate in the same way yourself. To do so, you have to learn about conventions for the use of notation, and also for the larger-scale layout of logical arguments. Such conventions might seem arbitrary, but conventions of any kind speed everything up: everyone knows what to expect, so everyone can process new information faster and more accurately. This is not unique to mathematics. If you were studying architecture, you would not do well unless you could construct and interpret standard diagrams and specification documents; if you were studying sociology, you would not do well if you could not construct a cogently-argued essay. If you want to be a mathematician, you will not do well unless you can present your arguments so that those in your field recognize them as coherent and logically sound. Adhering to standard conventions will help you to do that, and thus to gain recognition as a mature and intellectually-able mathematical thinker.

8.3 Writing a clear argument

In Chapters 5 and 6 we looked at a number of theorems and proofs, and I commented on things you should notice about them and incorporate into your own mathematical writing. If you follow that advice, you should not go too far wrong. However, certain errors are particularly common in undergraduates' writing. In this chapter, I will highlight some of these, starting with common errors people make when trying to present a logical argument.

First, consider the beginning of an argument. One thing students often do is write the desired theorem or conclusion as the opening line. For example, when addressing the question "Prove that if f is an even function, then df/dx is an odd function," they begin by writing:

f is even $\Rightarrow$ df/dx is odd.

I'm pretty sure that students do this because they want to remind themselves of what they are trying to prove, and in that sense it's sensible. But for a reader it is confusing, because it looks like the student has started by assuming the thing they are trying to prove. Of course, it's perfectly okay to write down what you want to prove, but you can help your reader by making the status of the statement clear. For example, you might write one of the following:

We want to prove that f is even $\Rightarrow$ df/dx is odd.

Claim: f is even $\Rightarrow$ df/dx is odd.

Either of these formulations shows that you know what you're going for, but also that you know you haven't proved it yet.

Second, consider the links between lines in an argument. In school mathematics, you will often have written things like this:

$$x^2 - 5x + 6 = 0$$

$$(x - 2)(x - 3) = 0$$

$$x = 2, 3.$$

Such a series of equations is not wrong, but it has no explicit logical connections between the lines. As a reader faced with something like this, I don't know whether the student means that each line implies the next one, or each line is equivalent to the next one, or something else entirely. This might not seem to matter but in fact it does, rather a lot. We can see this by looking at this set of equations instead:

$$\sqrt{x} + 2 = x$$

$$\sqrt{x} = x - 2$$

$$x = (x - 2)^2$$

$$x = x^2 - 4x + 4$$

$$0 = x^2 - 5x + 4$$

$$0 = (x - 4)(x - 1)$$

$$x = 4, 1.$$

Try substituting the solutions for x back into the original equation, and you will see that something has gone wrong. Can you work out where?

The answer becomes clearer if we use connectives properly. Some adjacent pairs of lines are equivalent to each other, but the step at which we square both sides is only valid as a one-way implication. While it is true that

$$\sqrt{x} = x - 2 \implies x = (x - 2)^2,$$

it is *not* true that

$$x = (x - 2)^2 \implies \sqrt{x} = x - 2.$$

So with appropriate logical connections, the argument looks like this (look at the arrows carefully):

$$\sqrt{x} + 2 = x$$
$$\Leftrightarrow \sqrt{x} = x - 2$$
$$\Rightarrow x = (x - 2)^2$$
$$\Leftrightarrow x = x^2 - 4x + 4$$
$$\Leftrightarrow 0 = x^2 - 5x + 4$$
$$\Leftrightarrow 0 = (x - 4)(x - 1)$$
$$\Leftrightarrow x = 4, 1.$$

So we can conclude that if $\sqrt{x} + 2 = x$ then $x = 4$ or 1, but we *cannot* immediately conclude that if $x = 4$ then $\sqrt{x} + 2 = x$, or that if $x = 1$ then $\sqrt{x} + 2 = x$. This is not a problem, because the real effect of this argument is to narrow down the solution possibilities. It tells us that the only possible solutions are 4 and 1, so then we just need to check whether either of them actually works.

Implication and equivalence arrows are often used in algebraic arguments. In other proofs, like many of those in Chapters 5 and 6, you will often see lines that begin with "Therefore," "Hence," "Thus," or "So." Any of these words indicates that the line after the word follows in a logically valid way from the previous line or lines, plus (perhaps) standard logical reasoning or algebraic manipulation, or a known definition or theorem. To avoid errors in using this type of language, one thing to note is that "so" in mathematics is always used to mean "so it follows that," *not* "so we need to prove that." Students sometimes forget this. They write something like:

We want to prove that f is even $\Rightarrow \mathrm{d}f/\mathrm{d}x$ is odd.
So $f(x) = f(-x) \Rightarrow f'(-x) = -f'(x)$.

What they mean is:

We want to prove that f is even $\Rightarrow \mathrm{d}f/\mathrm{d}x$ is odd.
So we want to prove that $f(x) = f(-x) \Rightarrow f'(-x) = -f'(x)$.

But what a mathematician reads is:

We want to prove that f is even $\Rightarrow \mathrm{d}f/\mathrm{d}x$ is odd.

So it follows that $f(x) = f(-x) \Rightarrow f'(-x) = -f'(x)$.

The last of these is confusing, because it looks like the student has started with what they want to prove and begun deducing things from it. This appears to a reader to be "backwards," because we actually want to start from the premise and work toward the conclusion (this was also discussed in Chapter 5).

The interpretation of "so" to mean "so it follows that" is just a matter of convention, and you are free to think that it is not the interpretation that you would choose. But, as I said above, part of becoming a mature thinker in any field is learning about its conventions and learning to adhere to them.

Now, what do you write if a new line does not follow logically from the previous lines? There are two situations in which this commonly happens. The first occurs when you need to introduce some object or notation. We can do this whenever we need to in a proof, and usually we do it either to introduce an arbitrary object or to explain what some notation will mean in the remainder of the proof. In such cases it is common to use the word "Let," as in:

Let $\theta \in \mathbb{R}$ be arbitrary.

Let $P(n)$ be the statement that $\displaystyle\sum_{i=1}^{n} i^2 = \frac{n(n+1)(2n+1)}{6}$.

The second situation occurs when a proof needs to develop two ideas separately, then put the results together to prove the theorem. I like the word "Now" for situations like this, and you will often see "Now" at the beginning of a line in the middle of a proof. What it means, essentially, is, "Right, we've just proved one thing we'll need, and now we're going to introduce something new before putting it all together." We often need to do something like this in longer arguments, and it is useful to have a way of indicating to the reader that we are directing their attention to a new idea. Such longer arguments appeared in Chapter 6 in particular, and it is probably worth looking back at some of those now, bearing in mind all of this information about conventional language.

8.4 Using notation correctly

This section discusses new notation that gets introduced in undergraduate mathematics, and highlights errors students often make while learning to use it. If you can manage to avoid these errors, you will make your professors very happy.

The first common error is easiest to illustrate using an example. Suppose you're trying to prove something about odd numbers. You know that, by definition, we can capture an odd number by saying

Let n be an odd number, so $n = 2k + 1$ for some $k \in \mathbb{Z}$.

But suppose that you're trying to prove something involving *two* odd numbers (that when we multiply them together, we get another one, or something like that). Students sometimes write things like

Let n and m be odd numbers, so $n = 2k + 1$ and $m = 2k + 1$.

Can you see why this is not appropriate? The definition has been used, but in sticking to it the student has inadvertently used the letter k for two numbers that might be different. This is easily fixed. You could write, instead,

Let n and m be odd numbers, so $n = 2k + 1$ for some $k \in \mathbb{Z}$ and $m = 2l + 1$ for some $l \in \mathbb{Z}$.

You might have noticed something like this in the proof about rational and irrational numbers in Section 6.3; that proof used p/q for one rational number and r/s for another, potentially different one.

The second common mistake is to use newly-introduced symbols when they are not appropriate. For instance, mathematicians use 0 for zero and $\emptyset$ for the empty set (the set containing no elements). Sometimes students learn the symbol $\emptyset$, but start using it when they mean 0. Now, zero and the empty set do have some similarities. Zero is a number, and when you add it to any other number you get the same number back again. The empty set is a set, and when you combine it in a certain way with any other set, you get the same set back again. Nonetheless, 0 and ∅ are not the same as objects: 0 is a number and ∅ is a set. Students sometimes get confused about this because they think of 0 as "nothing." But 0 isn't "nothing," it's a number; you could label it on the number line just like you could label any other number on the number line. So use "0" when you mean the number, and don't start using "∅" instead just because it's new.

Similar but even more common errors occur with the symbol "=" (equals). This amuses me, actually, because mathematics majors have been using the symbol "=" correctly for somewhere in excess of ten years, and probably only started misusing it in the last two. The misuse occurs when students use "=" not to say that two things are equal, but to act as a sort of placeholder between steps of a calculation. Here is a common example, in which the student is finding the third derivative of $y = 4x^5 + 2x$:

$$\frac{dy}{dx} = 20x^4 + 2 = 80x^3 = 240x^2.$$

This is just wrong. The first equality is valid, but the other two are not![1] Such a calculation should be written like this:

$$\frac{dy}{dx} = 20x^4 + 2, \quad \text{so} \quad \frac{d^2y}{dx^2} = 80x^3, \quad \text{so} \quad \frac{d^3y}{dx^3} = 240x^2.$$

This does involve a few more words and symbols, but it has the great advantage of being correct.

Students also sometimes stop using equals signs in places where they *would* be correct, because they have just learned the new symbol "⇔" and they're a bit over-excited about it. They start writing things like this:

$$x^3 + 4x^2 - 7x - 10 \iff (x+5)(x^2 - x - 2) \iff (x+5)(x+1)(x-2).$$

This is wrong for a more subtle reason. The reason is that "=" is for objects and "⇔" is for statements. The expression "$x^3 + 4x^2 + 3x - 10$" isn't a statement—we can't sensibly ask whether it is "true" or not, so it can't be

---

[1] The second and third equalities are invalid in general, at least—are there any specific numbers for which they do hold?

equivalent to anything. There are two ways to fix this problem. Presumably a student would be doing this type of calculation in order to find values of x for which the expression $x^3 + 4x^2 + 3x - 10$ is equal to 0. In that case, we could rewrite with equals signs, as appropriate for equating expressions:[2]

$$x^3 + 4x^2 - 7x - 10 = (x + 5)(x^2 - x - 2)$$
$$= (x + 5)(x + 1)(x - 2).$$
$$\text{So } x^3 + 4x^2 + 3x - 10 = 0 \iff x = -1 \text{ or } x = 2 \text{ or } x = -5.$$

Alternatively, we could rewrite the whole calculation in terms of statements that are equivalent to each other:

$$x^3 + 4x^2 - 7x - 10 = 0$$
$$\iff (x + 5)(x^2 - x - 2) = 0$$
$$\iff (x + 5)(x + 1)(x - 2) = 0$$
$$\iff x = -5 \text{ or } x = -1 \text{ or } x = 2.$$

Notice that for each of these better arguments, we can read out literally what is on the page because it all makes perfect sense in English. Try it. Now try reading out loud the original version, saying "if and only if" where you see the symbol "$\iff$". You'll find that you can't read it with sensible inflections; the sentence doesn't seem to "finish," because it isn't really a sentence. Mathematicians write in sentences. Sometimes the sentences have symbols in them, but they are sentences nevertheless.

Finally, this is probably a good place to say one more thing about accuracy in describing mathematics. Students tend to use the word *equation* to refer to anything with a lot of x's and other symbols in it. Technically, that's often incorrect. For instance, $x^3 + 4x^2 - 7x - 10 = (x + 5)(x^2 - x - 2)$ is an equation, because it's a statement about two things being equal. In contrast, $x^3 + 4x^2 - 7x - 10 < 0$ is an *inequality*. And $x^3 + 4x^2 + 3x - 10$ on its own is not an equation, because it doesn't say anything about equality. It should be referred to as an *expression* instead.

---

[2] Notice, by the way, that the "or" in the third line is important; it should not be replaced with "and" because we can't have $x = -1$ and $x = 2$ at the same time.

8.5 Arrows and brackets

Recall the example about the use of "$=$" in the context of derivative calculations. Can you see why the following common error is even worse?

$$\frac{dy}{dx} = 20x^4 + 2 \;\Rightarrow\; 80x^3 \;\Rightarrow\; 240x^2.$$

What would it mean to say that $20x^4 + 2$ implies $80x^3$? That's a bit like saying 2 implies 10 or 5 implies $\cos x$; it's meaningless because objects can't "imply" other objects. Students also sometimes write something similar with normal arrows, like this:

$$\frac{dy}{dx} = 20x^4 + 2 \;\rightarrow\; 80x^3 \;\rightarrow\; 240x^2.$$

In both cases, the arrows are not used in a mathematical way. They're really being used to mean something along the lines of "Yeah, like, and this is the calculation I did next." You might just about get away with this in some courses, but I would advise against it. Arrows often have specific meanings in mathematics, particularly in the definition of functions, and in courses involving limits (where "$\rightarrow$" can be read out loud as *tends to* or *converges to*). Thus, if you misuse arrows, you run the risk of writing something that makes no sense. Better advice would be that given by parents to their frustrated toddlers: *use your words*. If you are using arrows in a vague way, you almost certainly mean "so" or "implies" or "because of what we established in line (1)," so write something more precise instead.

Another common bugbear among professors is students' tendency to use the wrong kinds of brackets. Students can't really be blamed for this, because it hasn't mattered much before. If you just need to make the structure of an expression clearer, nobody cares whether you write

$$x((x + 1)^2 - 15x) \quad \text{or} \quad x[(x + 1)^2 - 15x].$$

However, during your major you will find that there are contexts in which brackets have quite different meanings. For instance, when we are talking about sets of real numbers, we use the following conventions:

- [0, 1] means the set containing all the numbers between 0 and 1, including the "endpoints" 0 and 1 (called a *closed interval*);
- (0, 1) means the set containing all the numbers between 0 and 1, excluding the endpoints (called an *open interval*);
- {0, 1} means the set containing just the numbers 0 and 1.

The first brackets are called square brackets, and the second are called round brackets, for obvious reasons. The last ones tend to get called "curly braces," which I, along with lots of students, find amusing for reasons I can't quite put my finger on. Anyway, notice that the square and round bracket notations only make sense if there are two numbers in the brackets, with the lower number first. When you see curly braces for sets, what appears inside is just a list, so it can explicitly include many more things, as in $\{1, 2, 3, i, 1 + i, 7.362\}$.

Curly braces are also used in set notation more generally, like this:

$$\{x \in \mathbb{R} | x^2 < 2\}.$$

This would be read out loud as "The set containing all the elements x in the reals such that x squared is less than 2." Sometimes people use a colon instead of the vertical line to mean "such that." But using round or square brackets in this case would be considered wrong.

Giant versions of curly braces are also used for piecewise-defined functions, as we first saw in Chapter 3:

$$f(x) = \begin{cases} x + 1 & \text{if } x < 0 \\ 1 & \text{if } 0 \leq x \leq 1 \\ x & \text{if } x > 1. \end{cases}$$

Notice that only the left-hand brace is used here.

There do remain cases in which bracket types don't matter much. You will still be fine with any type if you are just making clear the structure of an algebraic expression, and it is usually fine to write matrices with either round or square brackets:

$$\begin{pmatrix} 1 & 2 \\ 6 & -3 \end{pmatrix} \qquad \begin{bmatrix} 1 & 2 \\ 6 & -3 \end{bmatrix}.$$

However, you might find that professors use different bracket types, and if in doubt you should stick to the notation used in each course.

8.6 Exceptions and mistakes

With the exception of the section on brackets, I have been writing as though this advice about symbols and conventional layouts is set in stone. You should find that most of it applies everywhere, but you may find that one of your professors uses a symbol differently, perhaps because of the way it relates to the rest of the course. They might use "$\rightarrow$" instead of "$\Rightarrow$" to mean "implies," for instance, especially in a course that focuses on logic and doesn't have anything to do with limits. I wish we could clear this up, but I'm afraid that's not realistic: historically, different branches of mathematics developed somewhat separately, and everyone in each branch gets used to their own notation, so no-one wants to change it. This is annoying when you're a student, but it's something you have to get used to when working in a technical field. Just be alert to it and, if you're not sure, ask how a particular symbol is being used.

Also, remember that you are unlikely to write perfectly from the day you begin your major. Even if you have read this advice carefully, you will probably forget at least some of it when you are working on a complicated problem; when concentrating on a higher-level idea, you might not notice that you've made a notational error. That's fine. But sometimes it might be tempting to let things slide because it all seems a bit pedantic. Don't do that. Don't be lazy. Do it right. You can't claim to like the fact that mathematics has right answers, then also claim that you can't be bothered with this sort of detail. Being imprecise is sloppy. In some cases it will render your answer ambiguous; in some cases it will render it outright wrong. So, when someone points out such an error to you, the appropriate response is not "Oh, you're so fussy, you know what I meant!", it's "Oh, yes, I see the error and will fix it, thank you."

8.7 Separating out the task of writing

Hopefully you can see that good writing is worth aiming for. But you might find it difficult to imagine yourself being able to write everything well at the same time as doing all the necessary mathematical thinking. My advice about that is: don't try. You don't have to write everything nicely the first time, and it's perfectly acceptable to concentrate first on calculating a solution or finding a proof. But, after you've done that, you should look at your work with a critical eye and ask, "Now, could I improve the way I have written this?" In some cases, you might be able to improve it with the addition of just a few words. Perhaps a *Let* and a *Then*, plus a few *So*'s and a few commas and periods. A good test of whether you've done this well is to try reading it out loud—if you find yourself adding extra words of explanation, consider writing those in.

In other cases, it might be appropriate to add some justifications. Perhaps you could state some reasons at the beginnings or ends of some of the lines ("By theorem 3.1" or "Using the definition of continuity"), or perhaps you could include an extra calculation or argument to explain why a particular step is valid. In still other cases, you might want to rewrite things in a different order. Perhaps your proof would be better presented in the other direction so that it reflects the structure of the theorem. This kind of thing takes a bit of work, but only a bit. And you will find that writing appropriately becomes less arduous as you go on.

What I would say, though, is that you should always do this. Students often don't want to. Sometimes a student asks me to look at some work, and my response is that it's basically fine, but that the writing could be improved. The student often then says "But this is just scratch work." There are two counterarguments I make to this response. The first is that the more you practice writing well, the more naturally it will come to you, and the easier it will be to do it quickly in exams, and so forth. The second is that you will probably want to reread your scratch work in a few months' time when you are revising. Will you remember, by then, what you were thinking? Probably not. Writing a few words around it to indicate the reasoning will much improve the speed with which you can revise it.

The final thing I would say about writing mathematics well is this: hang in there. Every year I get a new set of undergraduate advisees, and every year, at the start of the year, their writing is some way from perfect. Every year, they go through a phase of using the new notation and language in a clunky and error-prone way, while they get used to it. And every year, at the end of the year, their writing is much better.

SUMMARY

- You can probably recognize good mathematical writing, even if you don't yet always produce it yourself. Professors will expect you to develop your mathematical writing skills.
- Writing well will earn you better grades on coursework assignments and exams, and will help you to be successful in professional life. It will also show that you understand how mathematicians communicate.
- When writing a mathematical argument you should make clear the status of the first line(s), use logical connectives appropriately, and adhere to conventional mathematical usage of words like "so," "thus," "let," and "now."
- Be careful with all symbols, especially new notation such as "∅" and "⇒". Don't be tempted to overuse symbols just because they are new.
- Different types of arrows and brackets have different meanings; make sure you know what these are so that you can use them accurately.
- It might be helpful to treat writing well as a separate task to be undertaken once you have worked out how to solve a problem. To decide whether your writing is good, try reading it out loud.
- You will almost certainly make mistakes when using notation or laying out logical arguments, especially when the mathematics is challenging. But most students improve their mathematical writing considerably during their major.

FURTHER READING

For more on clear mathematical writing, try:

- Vivaldi, F. (2011). *Mathematical Writing: An Undergraduate Course*. Online at http://www.maths.qmul.ac.uk/~fv/books/mw/mwbook.pdf.
- Houston, K. (2009). *How to Think Like a Mathematician*. Cambridge: Cambridge University Press.

If you are ready to consider more sophisticated aspects of mathematical writing, and you want to improve your written English while you're at it, try:

- Higham, N.J. (1998). *Handbook of Writing for the Mathematical Sciences*. Philadelphia, PA: Society for Industrial and Applied Mathematics.

PART 2
Study Skills

CHAPTER 9

Lectures

This chapter is about what to expect and what to think about when attending mathematics lectures. It contains advice about how to get the most out of being an independent learner in a lecture-based environment, and about how to deal with common problems.

9.1 What are lectures like?

Mathematics lectures have changed enormously in the 20-odd years since I was an undergraduate. Back then, all lectures were essentially the same. The professor would write on the blackboard and talk for 50 minutes, and the students would listen and write down what he or she wrote. Hopefully, the students would also use the professor's verbal comments to make extra notes so that they could understand everything when reviewing the material.

These days there is much more variety. Some professors still give chalk-and-talk lectures, although the chalk is often replaced with whiteboard pens. Some also provide students with outline notes or extra information via handouts that they distribute in lectures or make accessible on the university's virtual learning environment (VLE). Some use Powerpoint or overhead projectors instead of boards. Some put a full set of notes on the VLE in advance, then go over key aspects of these and/or work through more examples in lectures. Some put "gappy" notes on the VLE, and expect that students will come with their own copies and fill in the gaps during lecture time. Some professors do not make any notes available in advance, but do put them on the VLE after each lecture or at the end of each major section. I could go on—you will see lots of different approaches.

You will also see variations in patterns of professor and student activity. Some professors give an overview of the lecture before they start; some do not. Some give students breaks; some do not. Some expect students to spend quite a bit of time discussing mathematical ideas during lectures; some ask for this occasionally; some do not ask for it at all. Some expect students to contribute answers to the whole lecture class (especially if it is a small group); some do not. Again, I could go on.

9.2 What are professors like?

Still another source of variation is the fact that professors are human beings so they all have different personalities and skills. Some have upbeat and exuberant lecturing styles; others are more quiet and serious. Some are fairly clear (even dictatorial) about what should be written down and how; others leave that decision to the students. Some make a point of making eye contact with everyone around the room; some do not. Some have a good understanding of students' thinking and tend to anticipate likely mistakes; others are less experienced and sometimes speak at a level that is difficult to understand. Some have excellent, clear handwriting; some do not. You could probably have anticipated all that, but one thing you might not know is that it's likely that many of your professors will not be from your own country. Universities are very international places—they do their best to attract talented mathematicians from all over the world. In my department there are professors from Australia, Austria, China, Estonia, Germany, Greece, Italy, Lebanon, Mauritius, Mexico, New Zealand, Russia, Sweden, the UK, and the US, and that's fairly typical for a successful university. This is fantastic in many ways, but it does mean that some professors might have unfamiliar accents or styles of handwriting. You might have to listen and watch carefully until you adjust to each person.

The other thing about personalities is that you have one too, so you will respond to your professors in different ways. Some professors you will want to emulate because you admire them and enjoy their lectures. Others will have personalities that clash with your own, and you will not enjoy being in their company. Hopefully you'll have more of the former than the latter, but you'll have to make the best of all your lectures, which is what the next section is about.

9.3 Making lectures work for you

Amongst all this variety, there will be some things you like and some things you don't. Maybe you like interactive lectures given by enthusiastic types, but sometimes you find yourself totally passive in lectures given by someone you find very dull. If you're unlucky, this person might also have writing that is difficult to read and an accent that is difficult to understand. You will probably spend some time complaining to other students about this kind of thing, and you will probably enjoy it—complaining can be quite cathartic and it will serve to bond you with your peers. However, it isn't very constructive. No amount of whining to your friends will improve things so, after you've had some fun feeling hard done-by, you should consider formulating a plan to improve the situation.

The first thing to remember is that even if all your friends agree with you about the lectures in a particular course, other students might not. What you see as an incredibly dull presentation might seem, to someone else, magnificent in its clarity. What you see as a lively and enthusiastic presentation might seem, to others, to be scatty and all over the place. What you see as too slow a presentation might seem, to others, to be too fast (seriously, I sometimes get both comments on feedback forms for the same course). In lectures as in life, you have to get used to dealing with a certain amount of imperfection and to acknowledging that other people might have different preferences.

The second thing to remember is that if there is a simple, practical problem that would be easy to fix, it is much better to do something than to whine about it. First, use your common sense. If you can't see well, but that's because the room is huge and you sit right at the back with the cool kids, find some other venue in which to be cool and move forwards (or consider getting your eyes checked—apparently quite a few people realize that they need glasses when they start going to lectures). If, however, you've made reasonable adjustments, and you believe you would understand better if the professor spoke more slowly or more loudly, or wrote bigger or in a different pen, you can always drop them a quick email. Be brief and polite. For example:

Dear Professor X,

I am in your Linear Algebra course. Some of the other students and I are finding it difficult to read your writing. If possible, please would you write a bit bigger and in the black pen (that one is easiest to see)?

Many thanks,

Joe.

Do remember to specify which course you're talking about—professors sometimes teach more than one at a time.

This approach might also work if, say, the professor puts notes on the VLE, and gives extra notes in class, but does not always make it clear how the two fit together. They might be able to make the links more explicit via announcements or a numbering system, and they might be willing to do that. Again, though, make reasonable efforts yourself first—could it be that everything you need was explained in a course introduction document that you have long since forgotten about? Or that you simply haven't been studying very much and that an up-to-date student would have no problem following the presentation? If so, fix that first.

In any case, if you make requests like this, remember to be kind, because professors are people and they aren't necessarily very experienced at this part of their job. If you had to stand up and give three hour-long presentations a week about difficult, technical material, you probably wouldn't excel at it to begin with either. You might even be terrified, and you might respond well to someone telling you that they like some aspect of your lectures or lecture notes but would find it easier to learn from them if you made a minor adjustment.

9.4 Tackling common problems

Even with sensible adjustments made, you will still find that some lectures are not ideal for you. In that case, you should think about how you can learn as much as possible during lecture time and make sure you end up with a set of lecture notes that you can use effectively. Here are some suggestions that I would make in response to typical situations.

First, suppose the professor makes a whole set of notes available before the course starts, then spends the lectures going through these notes. You (organized student that you are) have printed these notes and have them in front of you but, because there's therefore nothing physical for you to do in lectures, you tend to get bored and drift off into a daydream. You might find yourself complaining that "the lectures are just the same as the notes" and claiming that there is nothing to learn in lectures. You will be wrong about this, of course—mathematics notes can be very terse, with all the key ideas included but without much explanation about how they link together or how an intelligent person should think about them (see Chapter 7). The professor will provide such information in their spoken commentary, so having the notes gives you an opportunity to make annotations that don't affect the basic material but that do capture how you should think about it. For instance, the professor might explain why one line in a proof follows from the previous line, or might link what you are studying now with something you learned earlier in the course. You could write down that sort of thing, and the professor probably expects you to do so. The professor probably also thinks that they have done you a big intellectual favor by allowing you to prepare in advance. Before each lecture, you could read through a roughly appropriate part of the notes, review any earlier concepts that seem to be needed, and highlight anything you don't understand so that you can pay particular attention when it comes up. This is part of what is meant by independent study—taking control of what you need to know and finding ways to seek it out. Of course, if that approach doesn't work, you could just stop printing out the lecture notes and force yourself to write everything down instead.

Second, suppose that the notes for a given course are not very structured. Perhaps there is not much of a numbering or sectioning system, so it appears that the material is just one long stream of definitions, theorems, proofs, examples, and calculations. In this case it might seem difficult for you to get to grips with how the course fits together. Again, though, there are some straightforward things you can to do help yourself. The first and most obvious thing is to politely ask the professor to introduce some sections. Perhaps, though, you could independently follow the advice given in Chapter 7, especially the suggestions about reading for synthesis: make a list of what's in your notes, consider numbering your pages, and perhaps turn the list into a concept map. Then you could impose your own numbering system, according to what seems to be the main breakdown of the course.

Third, suppose that the problems appearing in problem sets don't seem obviously related to the material in lecture notes. This is something that students complain about quite a lot. If you find yourself in this position, and if you haven't done so already, you should probably read Part 1 of this book. It might be that you are looking for worked examples when a professor is presenting material without many of these (see Chapter 1) and it might be that you can become more effective in reading (see Chapter 7) or in working out what to do when faced with particular problem types (see Chapters 1 and 6). You might also benefit from working with other students or from taking a list of questions to your professor or to another support person (see Chapter 10).

These are just examples of issues you might encounter and things you might like to do in response. You might come across other problems, and the important thing is to avoid getting carried away with blaming someone else—as I've said, you'll be expected to take responsibility for your own learning.

Speaking of taking responsibility for your own learning, another thing to be ready for is that you do not necessarily have to attend all your lectures. If you are working hard but finding a particular course very frustrating, and you think you might learn more if you just downloaded the notes and studied them independently, then maybe you should consider trying that for a week. You should do so with great caution and planning, however, and you should only consider this if you know you are a disciplined person who will do the work. There's more on why in the next section.

9.5 Learning in lectures

If you want to learn from a lecture, you have to go to it. I say this because colleges give you access to all sorts of interesting activities and people, and at some point you will find that one or another of these is more appealing to you than sitting still for an hour or more and listening to someone talk about linear approximations. However, very exceptional cases aside, you should go. One thing research is pretty clear on is that college students who consistently go to their lectures do better than students who don't.[1] Apart from anything else, going to lectures keeps you engaged in the process of being a student. You'll probably start off well in this respect—most people begin with good intentions—but do keep an eye on it. Don't use lectures you don't like as an excuse to slack off, and don't let missing a couple of lectures push you onto a slippery slope towards doing less and less work. If you do start slipping, catch yourself quickly and read Chapters 11 and 12.

Another thing to recognize about learning in lectures is that once you're there, you should pay attention. This, again, is not rocket science, but you might find it surprisingly difficult. In a lot of your lectures, the professor will talk for a full hour. Your high school teachers probably didn't do this; more likely they talked for a while and then gave you exercises to do, or let students work in small groups, or asked someone else to take over at the board. It might be hard for you to pay attention for a full lecture on upper-level mathematics, because the mathematical content will be more difficult than it was before, the pace will be faster, and you won't necessarily have a global sense of where it's all going. However, if you don't pay attention, you're not making very good use of your time. Things you don't think about in lectures, you're going to have to think about outside lectures, when you'd probably rather be getting involved with those interesting activities and people.

A third thing to recognize is that learning in lectures is easier and more enjoyable if you are reasonably well prepared. It's easier to understand new mathematics if you know the meanings of the key concepts in a course, if you're familiar with the types of argument typically used, and if you've studied the big theorems that glue the whole course together. Chapter 7 contains advice on how to get to grips with such things,

[1] If you have some knowledge of statistics, you might be aware that a correlation between lecture attendance and achievement does not necessarily imply that going to lectures *causes* people to do better—perhaps it's just that highly motivated people do well and happen to go to lectures too. But do you really want to take that chance?

and Chapter 11 contains advice on how to manage your time so that you can do so effectively.

A final thing to recognize about learning in lectures is that no-one is going to force you to do it. That's trivially true in one sense: where your high school teacher might have noticed when you'd drifted off and said, "Amy, come back to us please," your professor might not have a chance to learn your name and, even if they do, they almost certainly won't embarrass you by saying something like that in front of 200 other students. So you have more responsibility for keeping yourself on task. My point goes further than that, though. In most colleges, no-one is going to force you to go to lectures. A few colleges do keep track of attendance, but many do not. Indeed, in many cases, no-one is even going to notice if you don't go. If you've previously had a lot of attention from your teachers, it's easy to interpret this to mean that the people in your institution don't care about you. If you're a bit rebellious, it's also easy to feel like you're getting one over on the system if you don't go to lectures and you don't get into trouble for it. However, believing either of these things would be a serious error. If you don't study, you're not asserting your independence, you're just creating trouble for yourself. People do fail, and when they fail they have to do inconvenient and possibly expensive things like resitting courses. And your college and department *do* care about you. They would very much like you to succeed. But you're an adult now and, while they consider it their responsibility to provide high-quality teaching, they don't consider it their responsibility to make you study. That's down to you.

9.6 Courtesy in lectures

When you go to lectures, you should arrive on time. When the professor begins speaking, you should be ready: pen out, notes open at the appropriate page, and so on. You should turn off your phone. You should not speak when the professor is speaking. All of these things are common courtesy. They show courtesy to the professor, who has probably spent hours preparing this mathematical presentation for you. They also show courtesy to your fellow students, who have every right to be annoyed if you come in late, push past them to get to a seat, shuffle around getting stuff out of your bag, and then distract them by whispering to your friends. I'm going to expand just a bit on each of these points, because I know it's easy to feel anonymous in large lectures, and it's important to understand how apparently innocuous behavior might affect others.

In my experience, everyone knows how to be courteous, but people are often bad at acting on it. In particular, people are often late. I think this happens for two reasons. The first is that students often live very close to the location of their lectures. That sounds paradoxical but it isn't really—if you know it usually takes you 45 minutes to get somewhere, you tend to allow an hour, but if you know it only takes you five minutes, you tend to allow only five minutes. That means that if you forget anything or get distracted at all, you certainly will be late. The second reason is that people are accidentally a bit self-centered. They know that being late will disrupt their own education—they will miss part of a lecture, possibly the part where the main point gets explained. But they don't think about how their being late might affect others. Students coming in late puts the professor off their stride, and distracts the students who are already there and trying to concentrate. So pretend that the lecture starts five minutes earlier if that will get you there on time. If you do, you'll be relaxed and in the right frame of mind when you start trying to listen, which means that you'll get more out of lectures, too.

Now, cellphones. I would advise you to switch them off entirely. Mathematics lectures can be very intensive and learning from them is challenging, even for the well-prepared student. Cellphones and the like are sometimes known as "interruption technologies," for good reason. It's a bit daft to undermine your ability to concentrate by having something to hand that will constantly distract you. Also, be aware that when you use your phone, your professor can see you. From the professor's vantage point, it's very obvious when a student is sitting up straight, pen in hand, looking at the board, and paying attention. It's also very obvious when a student is sitting with their head down and one hand between their legs sending a text message. I once got a laugh by demonstrating this using a desk and chair that happened to be at the front of the lecture theatre. Sometimes people even forget themselves entirely, and sit with their elbow wedged on the desk and their phone in full view at the height of their head. Obviously that's even worse, but in either case it basically says to the professor, "I don't care how much effort you are putting into teaching me, I'm not even going to pretend to listen." Such behavior might not be intended as a personal insult, but that is how it comes across.

Finally, try not to talk in lectures. Again, I think that most people don't mean to be rude, but they get tempted into this because of the illusion of anonymity. They think that they can't be heard, or they think that there are so many students that the professor can't distinguish individuals, especially if they sit near the back. They are wrong, on both counts. I can always hear it when someone is speaking, and I can see everyone perfectly well. If you doubt this, pause at some point when you are walking across the front of a fairly full lecture theatre—you'll find that you can see all the faces quite clearly. Similarly, when someone is talking, I can usually tell exactly who it is. Even if you think your professor has poorer eyesight than me and is too nervous to tell you to be quiet, you should keep quiet anyway out of courtesy.

All of this said, there are occasions when you definitely *should* talk. If your professor asks you to discuss something with other students, you are being given an opportunity to articulate your current thinking about some idea, probably so that you have a good basis for development in the rest of the lecture. Take that opportunity. You should also raise your hand or otherwise politely speak up if you notice that the professor has made an error in their writing (writing an *n* when there should have been an *m*, or similar). This will happen sometimes—it's difficult for a professor to simultaneously speak coherently, write down a subset of what they're saying, and maintain reasonable eye contact with a roomful of students. Professors are grateful when people point out errors, because it's usually much easier to fix them immediately than to go back and do it later. Other students will be grateful too—if you've noticed an error, others probably have too, and at least some of them will be confused but too nervous to ask. If you're not sure whether there's an error or not, pop down and check with the professor afterwards (and see Chapter 10 for a discussion on other times and places at which you can ask such questions in more informal environments).

What I'm basically saying in this section is that I'm sure you are a well-mannered person who is considerate and courteous in most situations. Be like that in lectures too and it will be appreciated.

9.7 Feedback on lectures

A different sort of opportunity to be courteous arises when you are asked to give feedback on your courses. This will probably happen toward the end of each semester. Your professor will give out a standard feedback form, part of which will have a bunch of items where you mark "strongly agree," "agree," etc., and part of which will have more open questions ("What did you like about this course?"). When you get such a form, do read the instructions, because they are often machine-readable and, if you don't color in the little boxes correctly (or whatever), you will cause someone somewhere unnecessary work. Also, be careful with the "agree" responses. I once had two students run up to me after handing in their forms and say, "We're really sorry! We read the numbers the wrong way around and it looks like we hate you when actually we think you're really good!"

More importantly, though, please take the opportunity to write constructive comments. I stress, *constructive* please. There really isn't much point in responding to a question like "What did you like about this course?" by writing "Nothing." This comment contains no information that would help a professor to improve. Try to be specific and, if you have a complaint, give a constructive suggestion about what could usefully be changed. Make sure it realistically can be changed, of course—a person probably can't stop having a Chinese accent, but they might be able to speak more slowly. This request about specificity also applies if your comments are positive. As a professor, it's lovely to get positive feedback, but it's also useful to get specific information about which aspects of the course best supported the students' learning.

You might have more informal opportunities to give feedback, too. A professor might give out a questionnaire early on, to get a sense of what is and is not working. Or they might ask for feedback on an activity or resource that they're trying out for the first time. Again, be constructive in your responses, and make polite requests for small adjustments if these would help. Finally, if you enjoy someone's lectures and learn a lot from them, do find a way to let them know that. Professors often put a lot of work into their courses and it's nice to have this acknowledged. You don't need to go overboard—a simple "thank you" on the way out of a lecture will be much appreciated.

SUMMARY

- Lectures vary a lot in terms of presentation medium, handout provision, and the degree to which students are expected to interact.
- Professors have different styles, some of which you will like more than others. Many professors are from overseas, so you might have to adjust to different accents and styles of handwriting.
- If you have a small practical problem with a course, think about whether you could fix it via your own behavior. If not, make a polite and constructive request for a change.
- You might find it difficult to concentrate if lecture notes are give out in advance, or difficult to see how a course is structured, or difficult to link course notes to problem sets. There are sensible ways to tackle these issues.
- You should go to lectures and pay attention. This is your responsibility, and it will be easier if you make some effort to be prepared.
- Courtesy in lectures will be appreciated: be on time, switch off your cell-phone, and do not talk when the professor is talking.
- When you get opportunities to give feedback, do so in a specific and constructive way.

FURTHER READING

For more on learning from lectures and lecture notes, try:

- Moore, S. & Murphy, S. (2005). *How to be a Student: 100 Great Ideas and Practical Habits for Students Everywhere*. Maidenhead: Open University Press.
- Newport, C. (2007). *How to Become a Straight-A Student: The Unconventional Strategies Real College Students Use to Score High While Studying Less*. New York: Three Rivers Press.

CHAPTER 10

Other People

This chapter gives suggestions on how to get the most out of interactions with professors and other instructors, and on things to consider when working with other students. It also discusses opportunities you might have to interact on a more individual level with project supervisors and internship managers, and describes common college support services.

10.1 Professors as teachers

You will learn a lot from other people during your undergraduate education. This is obvious in the sense that your professors are supposed to teach you, but there are also many other ways to interact with different people during your studies. Toward the end of the chapter, I'll discuss studying with other students, but I will begin with opportunities you have to interact with professors and others in authority.

At college it might not be obvious that you have access to authoritative people. In high school, your teachers probably offered you help on a daily basis. They talked to you individually, noticed when you were having trouble getting started on a problem, that kind of thing. In college, classes can be bigger, so your professors probably won't do that. In a class of 100 people, it just isn't practical for a professor to catch someone and say "So, Nina, how did you get on with yesterday's work?"

Because of this, it's easy to feel distanced from your professors, and to get the impression that there is not much support available. This is not the case at all. There is support—lots of it—the difference is that you have to take the initiative and seek it out. This might initially seem daunting, especially if you are a bit shy. Professors tend to look very authoritative and hyper-intelligent, and some people find it intimidating to talk to them. But they are just people, and they love mathematics, and the vast majority of the time they will be delighted to talk to you about it. In this chapter I'll talk about how to have good, effective, individual interactions with your college teachers.

10.2 Recitations and problems classes

We'll start with recitations (sometimes called problems sessions) because these might be the most obvious place to get your mathematical questions addressed. As I said in the introduction to the book, many courses have associated recitations, and these might be run by the course professor or by another instructor such as postgraduate teaching assistant who is studying for a PhD in mathematics. Recitation sessions are often smaller than lecture classes, which means that recitation instructors might be better placed than professors to get to know you personally and to answer your particular questions.

What happens in recitations varies from institution to institution and from instructor to instructor. You might be expected to work on an assigned set of problems every week and to hand in written solutions or to bring these to the session. However, it might be much freer than that. Your instructor might expect you to take the initiative and bring along problems that you got stuck with or things from lectures that you didn't understand. In that case, the effectiveness of the sessions will depend on how well you prepare.[1] Ideally, you should take along a list of things you want to ask, as well as all the relevant notes and problems and your own solution attempts.

Who does what during recitations varies too. Your instructor might do most of the talking; they might ask what problems the students are stuck with, then work through solutions at the board. Alternatively, students might be expected to work in small groups while the instructor goes around to help people when they get stuck. In that case, you'll get more

---

[1] I've heard of an instructor who sends his students away again if they show up without questions. That sounds a bit extreme to me, but I understand his position. The main responsibility for learning at college lies with the student.

individual or small-group attention, and probably therefore some input on how to improve your individual thinking and writing. Your instructor might even sit at the back and expect students to do most of the writing and talking, with hints from them as appropriate. You should embrace this if it happens—it's probably the best opportunity you will get to articulate your thinking and to receive guidance on how to improve.

However your recitations are organized, you should feel free to ask about things other than the problems that the instructor is discussing. I don't mean interrupt if they're talking to the whole class, obviously, but you might be able to email in advance and ask them to go over a particular problem or to re-explain an important concept. If there's something you didn't understand in a lecture, then hearing an alternative explanation from an instructor might be just what you need. This process should become more effective as time goes on—as your instructor gets to know the class, they will develop a feel for what the students already know and thus for what explanations are likely to make the most sense. And if they do come around to give individual help, that's a good time to ask questions about other parts of the course too.

Overall, your recitation instructor is well-placed to support you as an individual mathematical thinker and to help you to improve your on-the-spot reasoning, especially if they also give feedback on your written work. This means that you should always go to recitations and make the most of what is on offer. If your instructor is a professor, there is also another, more prosaic reason you should always go: at some point you might want this person to write a recommendation for you. If you ask them to do so and their first thought is, "Oh yes, Martin, he's the one who rarely showed up and was always ill-prepared," that doesn't put them in a great frame of mind for writing about how brilliant you are.

10.3 Asking questions after and before lectures

You can also put questions directly to your module professors in lecture time. However, most people are not comfortable asking questions in lectures, especially in big lower-level classes. This is completely understandable—there's always a risk that you've misunderstood something and, while it's good to develop confidence in voicing your own uncertainty, it's normal to avoid taking that risk in front of 100 other people. However, there's nothing to stop you going up at the end of the lecture and asking your question then. This is perfectly acceptable, and is something that many professors actively encourage. I certainly wish that more students would do it, not least because people who do ask questions often do me a big favor. Sometimes it turns out that they're confused because I made a small error that no-one pointed out at the time; knowing this allows me to correct it in the next lecture. And sometimes it turns out that they've misinterpreted something I've written or said. Since people don't usually make crazy interpretations, other students might well be confused for the same reason, and again I can use this information to give an improved or different explanation in another lecture.

Of course, asking questions at the end is probably only practical for small points of clarification. If you're properly confused, you might not be able to formulate your question that quickly, and you might need to go away and think about it for a while. In that case, you might want to arrive early for the next lecture and ask before it starts. That's fine too. Bear in mind, though, that your professor might not be able to focus on your query straight away. They might need to concentrate on setting up for the lecture, or they might have to rush off at the end because they've got a meeting on the other side of campus. Even if they can concentrate on it, they might decide that they need more than a couple of minutes to give you a proper answer. In either case, they might ask you to come back another time or to make a separate appointment with them. Don't feel put off by this—it's usually just because of practical considerations.

10.4 Arranging a separate meeting with a professor

In some cases, you might decide that a longer meeting with a professor would be useful. Perhaps you are having trouble even formulating what is confusing you, or perhaps you have built up a list of questions and it seems sensible to go and ask them all at once. Professors usually have assigned office hours for this kind of thing. Office hours are publicized in module documentation or on the web, and you can just show up at the stated time and place (usually the professor's office). If a professor doesn't have office hours, don't let that put you off. It's perfectly acceptable to drop a professor an email and request a meeting with them. If you want to do that, here are some suggestions for making the process effective.

First, suggest a day on which you could come (usually you'd go to the professor's office in these circumstances). If I were you, I'd say something like "Perhaps I could come on Tuesday afternoon? I am free 2–4pm if you have any time then." I wouldn't be any more specific than that—I find it a bit irritating when students say things like "I would like to come and see you at 3:30 on Friday," because it's presumptuous for them to act like I'll just be able to drop everything and meet whenever they like. Perhaps I'm atypically irritable in this respect but, however you do it, show some flexibility—professors are busy people (see Chapter 14).

Second, consider discussing your questions with a small number of other students first, and arranging to go along together. Your professor will appreciate this, as it makes for an efficient use of time.

Third, make sure you're clear in advance about what you want to ask. It's quite hard to help someone who comes in with a vague sense that they "don't understand Chapter 3;" it's much easier to get to the heart of the difficulty if they say, "I've got up to here when working on problem 2 but I don't know what to do next," or "I understand most of this proof but I don't know why the step from the third to the fourth line is valid." Take along a list of questions to remind yourself exactly what you wanted to ask and to make sure that you don't miss anything. And don't worry if the list is quite long. I've noticed that students can be embarrassed to show up with a long list, but I love it—it shows that they've been organized, and it usually means that we can handle a lot of queries in a pretty short time. Of course, if you genuinely are mystified by a whole section of a course, then it's fine to ask, but try to frame your question constructively. You could say something like, "I was okay with most of Section 4, but I've been struggling to follow Section 5, and I wonder if I've missed a key idea that would make it all easier." Then at least the person has some sense of where to start.

Fourth, take along everything you might need to refer to. This is important because your professor might not have everything immediately to hand, and because they don't have encyclopedic knowledge of exactly what question 5 from problem set 6 was about (I'm sometimes touched by the faith students have in my memory of exactly what I've written or asked, but usually this is misguided). In fact, it can be useful to have your version of everything for another reason: sometimes I've looked at a student's notes and discovered that the reason they're confused is that they've mis-copied something. And it goes without saying that you should make sure you've got all your materials organized so that you can easily find the things that are relevant to each of your questions.

Finally, remember that it is part of your professor's job to support you in the course, so you should feel perfectly justified in going to see them. This does not mean acting like a 6 year-old, of course. You shouldn't knock on their door the minute you can't do something—obviously you're expected to give it some proper thought first. But, if you do decide you want to see them, you might find it a surprisingly pleasant experience. As a student, I found this out much later than would have been ideal, when I went to see a professor toward the end of a masters course. He was very happy that someone was taking an interest in his subject, and he was willing to talk to me for ages.

10.5 Asking questions electronically

These days, I receive quite a few mathematics questions by email. If I'm honest, getting these makes me feel a bit weary. I'm glad the student is asking, obviously—much better that they ask a question than worry about something or get confused and stay that way. But, even with the many recent improvements in technology, it can still be hard to deal with mathematics in an ordinary email environment. Sometimes I find it difficult to understand what exactly the student is asking. Other times, I can understand what they're asking, but I can't formulate a very clear response in writing. Usually, I feel it would be much better if the student was actually there in the room, so that we could look at something on paper and have a back-and-forth discussion in which I could try to make sure we understood each other. The upshot of this is that when someone emails me a mathematics question, I often end up asking them to come and see me instead.

That said, I do know other professors who are keen to communicate with students using various technological tools, and who make use of things like discussion threads on course webpages. This is another situation in which you should think about what is practical and be alert to what a professor seems to prefer.

One thing to remember, though, is that whenever you contact a professor, you should write properly, with (as far as you can manage it) correct spelling and grammar and punctuation. This is because you are in a professional relationship with this person—indeed, they are your senior. They might act in a relaxed and informal way in class—they might tell you to call them by their first name and so on—but they are not your buddy. Make sure you write a proper greeting ("Dear Lara" or "Dear Dr. Smith"), be polite, and do not write in textspeak (few things make you look more immature than writing "I wud like 2 c u" or similar). Make sure you sign off with your name so the person knows how to address you in return, and for heaven's sake don't put kisses after it. And, in the long term, if you know that your written English is not very good, consider getting a short book and learning how to fix it (see the Further Reading section for suggestions). One thing that really puts off employers of all kinds is a person who doesn't know where to put their apostrophes or who makes other glaring grammatical errors. You might as well start learning to fix this now.

10.6 (Mathematics) learning centers

Support services in colleges tend to expand and get more specialized as time goes on, and lots of of institutions now have *learning centers* (sometimes called *tutoring centers* or similar), some of which offer specialist mathematics tutoring. Both general and specialist centers are usually open to students from across the whole institution—lots of degree programs require both general learning skills and one or another kind of mathematics—although some might specialize in helping either newly-arrived students or majors from particular departments. Don't let the name make you think that learning centers are only for people who are really struggling—usually they are open to people who are doing well too.

There are a couple of basic models for learning centers. In some institutions, you book an appointment (probably online), and then show up at the appointed time and get one-to-one help. The person helping you might be an instructor or it might be an undergraduate or postgraduate student. Other learning centers operate as open study spaces, where you can just walk in and join a queuing system. In my university this involves picking up a number on a laminated card and putting it in front of you on the desk, so that the tutor on duty knows that you would like them to come to you. Often, learning centers have various mathematical resources (books, computers, leaflets) around too, and often it's fine for you to just go in and use them as places to study, either on your own or with other students.

Indeed, learning centers that operate as study spaces can facilitate cooperation between students. In the Mathematics Learning Support Centres at my university, I often see groups of students working quietly together. I have also seen acts of genuine friendship and support between students who did not even know each other. Once, I was trying to help someone with an unfamiliar upper-level course, and I wasn't doing a great job because I didn't have the right knowledge at my fingertips. I was rescued from this situation when a student came over from another table, pointed back at her friends and said "We're working on that over here and I think we can do that question—you're welcome to come and join us if you like." What a lovely moment.

One thing to bear in mind is that, in a learning center, you might get help from a tutor who knows all about the subject you are studying. This should almost always be the case if it's a specialist center and you want help with a lower-level course. But you are now reaching the stage of your mathematical education at which people start to specialize, and it might be that the tutors do not offer help with upper-level courses. Some tutors might still be able to help you—they might have studied similar material, and they might have better general problem solving skills than you do— but they will probably need to check some definitions or to find out how your course professor is using a certain notation. So think about where best to get help, and, as ever, take along all the relevant materials.

Another thing to bear in mind is that tutors in a learning center might be qualified to give you other types of academic guidance, such as advice on handling learning difficulties, or on which courses to take at which stages of your degree program. But they might not. They might be students who are able to help with certain mathematics courses but who are not qualified to advise on the college's liberal arts requirements, for instance. For that type of advice you should always go to a qualified person; advice from other students, however well-intentioned, is often out-of-date or otherwise misleading.

As you can probably tell, I think that you will have lots of opportunities to develop your understanding by talking to more experienced people. However, despite all these positive comments it is possible that, at some point, you (or a friend) will have a bad experience when asking for help. This is very, very rare, but I do occasionally hear that an instructor or tutor has made a student feel bad by telling them they should already know how to do something, or has acted surprised at their lack of knowledge. In the unlikely event that this happens to you, the thing to remember is that this person has behaved insensitively, and that you don't want to let one insensitive person (or perhaps an ordinarily sensitive person who's just having a really bad day for some reason) put you off. So complain about the experience to a friend, then put it behind you and ask someone else.

10.7 Projects and internships

So far, this chapter has discussed ways in which you might interact with professors or with other experienced people during day-to-day undergraduate activities. This is probably all you will experience during your first year or two, when you will go to lectures, seminars, and recitations. However, later in your degree program you might have the opportunity to take part in activities that involve different types of interactions and that develop different skills.

First, you might have the opportunity to work on an extended mathematical essay, project, or dissertation. This might occur in your final year, where it would act as a replacement for one, two, or perhaps more standard lectured courses. Or it might occur as a summer research opportunity, especially if you are in an honors program. Now, projects (I'll use that as an umbrella term) have some risks attached. By the time of your final year, you will have a pretty good idea of how well you can do in exams, whereas a project is a bit of an unknown quantity. However, that very fact is also key to the advantage of doing a project: it will develop different skills. This is a point worth serious consideration. Mathematics courses, more than many others, are pretty uniform in that students mostly learn from lectures. Mathematics majors don't usually do a great deal of work in groups, for instance, or give a lot of presentations, or write reports and critique those written by others, or independently research some topic. This situation isn't bad in itself, but it does mean that mathematics majors can lack experience to draw on when potential employers want examples of how, for instance, they have developed their communication skills.

If you do a project, you will get experience that addresses at least some of these points. You will learn at your own pace from books or research papers, or you will work on applying a standard argument to a new example, or something like that. You might have to do a presentation to other students and/or professors, either part way through or at the end. You will almost certainly interact one-to-one with the professor who is acting as your project adviser; this person might also give you periodic feedback on your written work. This can be especially valuable if you think you might want to go on to further study, as it's much more like being a PhD student than is sitting in lectures. In addition, you might find that you really enjoy the experience and learn a lot from it. There is something inherently satisfying about producing a substantial piece of written work that reflects a deep understanding of a topic, showing that you can structure a discussion of it in a way that is coherent and clear. So, while I can't comment on any specific situation, my view is that the benefits of doing a project generally outweigh the risks.

Another opportunity you might get is to go and spend a semester or a year studying abroad. Usually you can't choose to go just anywhere (there's no "I'll take Paris, please!") because there has to be comparability between courses you would have taken at your college and courses you will take while you are away. But your institution might have links with a number of universities around the world, and you might want to look into what is available. This is especially true if you are studying a language or if you intend to do so. But you might also find that there are links with universities in the UK or in other English-speaking countries. In any case, going abroad for a period of time will provide you with experience of a different educational system, and will allow you to get to know both students and professors who come from a different cultural background. Again, this has obvious benefits that will be recognizable to future employers, as well as being enjoyable for its own sake and likely to improve your general level of self-confidence.

Finally, you might have the opportunity to do an internship. This is not so academic: it typically refers to a period of time spent working as a (paid or unpaid) assistant in a business of some kind. Lots of major companies and other organizations offer such internships, and students often take them up during a summer vacation. Typically, your careers service or department would be able to provide you with general information on internship applications and on how to find vacancies, and then you would take the initiative to apply and go for interviews, just as if you were applying for a longer-term position. If you are studying for a minor or double major that has an obvious business or finance or economics component, it might be that everyone on your program tries to get an internship. If not, it probably won't be the case that everyone does it, but you might still have the opportunity. I would suggest looking into the possibility, at least if you want to go into a career in finance or in business (interpreted broadly). An internship can allow you to learn a great deal about your chosen industry; to pick the brains of those who work in it and to develop the sort of business acumen that it's difficult to get by sitting in theoretical courses. In my experience, all students who work as interns get an improved understanding of what it's like to work in that particular sector, and some of them get job offers out of it. That's not a trivial thing when you consider the practicalities—you can focus much better on the final year of your studies when you know you're already set up with a good job.

In all of these project and internship situations, your usual pattern of relationships with other people will be shaken up. You will not be just one student among many; instead you will interact closely with particular people, many of whom you would not otherwise have met. Some students welcome this type of variety and change in life, while others don't, and I'm not saying that everyone would benefit equally. But you might want to find out early on what is on offer, especially as some programs require you to maintain a certain average grade if you want to take part. As with other issues raised in this book, I'd just advise you to be well-informed and to make your decisions thoughtfully.

10.8 Studying with other students

For now, though, back to the usual business of undergraduate study, and in particular to studying with other students. Other students are plentiful, they are all at least reasonably clever, and lots of them are nice. This makes them an excellent resource to draw on when learning new mathematics, and doing so can make learning more enjoyable.

Depending on your previous experience, you might already be accustomed to working with other students. I often see students sitting down together with notes spread out everywhere, sharing ideas and discussing how best to do things. I'm not sure how the groups find each other in the first place, but talking to people in lectures and recitations will certainly provide you with a way to get a study group off the ground. Once it is off the ground, of course, you should think about how well it's working. On the one hand, having an agreement to get together and work on something can make sure you actually do it, a bit like having a buddy for going to the gym. On the other hand, if a bunch of you get together with the aim of working on mathematics, but actually you just distract each other, that's probably not very useful. It might be worth considering a middle ground in which you all agree to look at something in advance and then get together to discuss parts you were stuck with. You should also keep an eye on what you are getting out of any particular discussion. I wouldn't worry too much if your friend is better at a particular course than you; they'll improve their understanding further by explaining their thinking. But I would worry if you can't get a word in edgeways, or if you end up just copying down what someone has said without understanding it. In that case, you're probably not doing yourself any long-term favors.

Alternatively, you might not like the idea of working with others. You might find that you can concentrate better when you're on your own, or you might just think of mathematics as a solitary pursuit. This is also fine, and in most cases no-one is going to force you to work with other students (at least in lectures and independent study—some instructors might expect you to work with others during recitations). If I were you, though, I would think about getting the right balance, and make sure you're not working in seclusion just because you think that's how genius mathematicians do it. Genius mathematicians do sometimes work alone, but they also sometimes work in research groups, and either way they certainly engage with wider networks of mathematicians by going to conferences and seminars and debating whether a proof is valid or whether an idea will really generalize to a new type of object. Whether or not you want to be a mathematician yourself, it's a good idea to practice taking part in mathematical discussions and debates.

Also, bear in mind that working with other students can calibrate your sense of how well you are doing. If you study in isolation, you might end up with a rather inaccurate sense of how your understanding compares with that of other students. That might be fine but, if you feel you're not doing very well, isolation can easily exacerbate this (see Chapter 13). So, if you find yourself struggling on your own, consider spending at least some of your study time with others. You will probably find that you are doing better than you think.

It probably sounds like I'm saying that students should work together a lot, but I'm not sure I'd actually go that far. When I was an undergraduate, I spent a lot of time studying alone, and obviously that turned out well for me. But, with hindsight, I worked that way at least partly because it was what I was used to—at high school in the UK I was the only person taking something called Further Mathematics, so there weren't really other students I could talk to, and I never got into the habit. Later in my degree program, I did study more with others. I found that doing so improved my confidence, and that I really enjoyed it and wished I had done it before. Of course, if you are a part-time student, or a mature student, or if you live at home instead of on campus, you might find it more difficult to get together with other students for purely practical reasons. But most people have at least some breaks between lectures during the week, and you could have a regular study session in one of those.

10.9 Support with everything else

While you're an undergraduate, you won't just be a mathematics major, you'll be a whole person with a whole life. Hopefully that life will be excellent, and you'll make progress toward a fulfilling understanding of mathematics, a good grade point average, and an interesting career in whatever you want to pursue. But four years is a long time, and occasionally things do go awry for reasons that are beyond your control. You will get information about support services when you arrive at college, but I'll say a bit about those that are typically provided.

I'll start with careers services. College careers services are, in my experience, staffed by well-informed and helpful people who are just dying to help students get into interesting and well-paid jobs. Their biggest problem is that students don't get around to going to see them, or at least they don't get around to it until considerably later than would have been ideal. I can see why this happens—for mathematics majors at least, there is always a problem set to work on or a test to prepare for or another lecture to go to, and these things always seem more urgent than thinking about the future. Also, the future is usually not well mapped-out and is thus a bit scary, and people are sometimes inclined to avoid thinking about it.

But it's really not hard to make effective use of a careers service, even if you haven't a clue what you want to do with your life (which is very common in people starting a degree program). If you are in that situation, I'd suggest that you go to the careers service at the earliest opportunity, tell someone behind a help desk that you intend to major in mathematics, say that you don't know what you want to do in the long term, and ask what literature they can give you to help you be better informed about careers that might be open to you. They will load you up with nice, magazine-like glossy brochures about jobs in a variety of different sectors, which you can peruse at your leisure. When you've done that and you've got a better sense of what might interest you, you can go back and make an appointment to get more detailed advice. If, on the other hand, you already have some idea of what you'd like to try, you can make an appointment early on and ask for advice on how to, say, get a summer internship. If you can't even face any of that, at least look out for emails from the careers service telling you about employer presentations and about careers events in which lots of employers send along (usually young) staff to tell students what their jobs are like. Go to this kind of thing and you'll get a sense of what's out there.

Again, help is there but you have to be active in seeking it out. Not *very* active—it's not much effort to do what I just suggested. But don't wait for someone to tell you to do it. They might—if you happen to meet an adviser like me, you might have someone nagging you about it reasonably regularly. But a lot of mathematics faculty don't consider this part of their job. They'll assume that you're responsible enough to take care of it for yourself. Some students do, but others don't, and those who don't end up trying to find out about jobs while studying for their final exams. If you do some information-gathering early on, you can avoid getting into this position.

In addition to a careers service, you will also find that your university has support for other practical things. Obviously there will be a library. There might be some kind of printing centre, used by students for binding big project reports. There will be a central student administration office, where you might go to do things like register and pay fees. There might be a student job center to help those who want to get part-time jobs in the local area. There will be an accommodation office, which will help students to arrange on- or off-campus housing. There might be a computer support center, and there will probably be a medical center—make sure you find out how to register if you want to use it. There might be a general practical advice service, where you can go for assistance if you run into any problems with your landlord or the electricity company or whoever. There will also be an academic advice service to help students make good decisions about their degree programs. Finally, your department itself will probably have a main office, where you might go to hand in coursework assignments or to get advice specific to your major.

There also will be some sort of disabilities and additional needs service. If you are dyslexic or have a movement or sensory impairment or something like that, get in touch with them, because they'll be able to get you any support you need to study effectively. If you think you might have such a condition but you've never had it diagnosed, they can probably get you properly tested. If you are in such a situation, I would also make a plea with you as a professor. Services like this keep their students' needs confidential, which means that I don't automatically get told if there is someone in one of my lectures who needs something. Sometimes, for instance, a note-taker shows up for a student who is dyslexic or who has a visual impairment, but I have no idea who the student is. Sometimes I could do something simple to help this person—write in a particular color, or provide a larger-sized set of notes. But that information won't reach me very quickly. So, if you are in such a position, and there is something simple a professor could do, ask them if they'd be willing to do it. If it is practical (and fair to everyone else, of course), they will probably say yes straight away, and you might find this gets you what you need more quickly than going only through official channels.

Hopefully these practical services will be all you need during your time as a student. But there are services around for emotional and psychological support too. Sometimes people need these services for obviously serious reasons such as experiencing a severe illness, or a death in the family, or an unplanned pregnancy, or an addiction problem, or an abusive relationship, or depression, or another mental health issue of some kind. Such things, mercifully, are rare, but they do occur from time to time. Sometimes people need the services for reasons that are less obviously serious but that can nonetheless interfere with their ability to study. Perhaps they break up with a long-term boyfriend or girlfriend, or they realize they are gay, or their parents decide to get divorced, or they get in with a crowd that's drinking too much, or they are genuinely worried about failing their degree, or they are just much more homesick than they were anticipating. In any of these cases, getting expert support might be really useful.

Support services should be easy for you to access. If you live on campus, you will have a resident assistant (RA), who should be a good first port of call if something is bothering you. Whether you live on campus or not, you will have access to a college counseling service. Information about all of these services will be provided when you arrive and will likely be accessible from the page that first opens when you access your institution's website.

What I would say is that if you do have a problem like any of these, please go to one of the services to get help. I appreciate that people like to solve their own problems, but you do not have to feel ashamed, or to struggle on by yourself. Colleges are very accustomed to dealing with such issues, and no-one will be surprised by what you are saying. This all applies, also, if you are a friend to someone who experiences a problem. It's easy to feel that you have no right to be upset when your friend is the one having a bad time, and obviously you'll want to focus on supporting them. But sometimes in order to do that you need to get a bit of advice or support yourself.

This all sounds a bit depressing, so I want to end by saying that the vast majority of students have a great time at college and never have any of these problems or even meet anyone who has. They make many new friends and they enjoy positive relationships with their professors and other instructors. So you shouldn't worry about anything in advance. Just remember that if you do have a problem at some point, support will be available.

SUMMARY

- Professors and instructors can be accessed in lectures, recitations, and sometimes (mathematics) learning centers; students are expected to take the initiative in seeking help.
- Whenever you ask questions, it helps to have all the relevant materials with you and to be clear about what you want to know. Take a list if appropriate, and make sure your materials are well-organized.
- Think about when would be the most practical time to get your questions answered and, if you send an email, be polite and write well.
- Find out what your college supports by way of projects and internships. These can help you to develop your confidence, to gain different skills that are valued by employers, and to have more individual relationships with senior people.
- Consider your strategy for working with other students. Doing so can make learning more enjoyable and can improve everyone's skills in talking about mathematics, but make sure that you are not just getting distracted.
- Colleges have many support services. Everyone should go to the careers service early. Hopefully you won't need to use the other services very much, but it is probably worth familiarizing yourself with how they work.

FURTHER READING

For more on interacting with others and managing the practical aspects of college life, try:

- Newport, C. (2007). *How to Become a Straight-A Student: The Unconventional Strategies Real College Students Use to Score High While Studying Less.* New York: Three Rivers Press.

For more on career options for mathematics majors, try:

- http://www.maa.org/careers/
- http://www.ams.org/careers/
- http://www.siam.org/careers/thinking.php

To improve your written English, try:

- Seely, J. (2004). *Oxford A–Z of Grammar & Punctuation*. Oxford: Oxford University Press.
- Strunk Jr, W. & White, E. B. (1999). *The Elements of Style (4th Edition)*. Longman.
- Trask, R. L. (1997). *The Penguin Guide to Punctuation*. London: Penguin.
- Trask, R. L. (2002). *Mind the Gaffe: The Penguin Guide to Common Errors in English*. London: Penguin Books Ltd.

If you are writing a larger mathematical project, try:

- Higham, N. J. (1998). *Handbook of Writing for the Mathematical Sciences*. Philadelphia: Society for Industrial and Applied Mathematics.

CHAPTER 11

Time Management

This chapter is about how to organize your study time so that you can keep up with your courses and avoid the stress associated with falling behind. It offers practical suggestions for making realistic plans that you can actually stick to, and for factoring in social activities so that you can enjoy these without feeling that you should be studying.

11.1 Why would a good student read this chapter?

Presumably you are a good student already, so you might doubt that anyone has anything to teach you on the subject of time management. In that case, you might like to know that every year, when advisees come back, I ask them whether they will do anything differently. Every one of them says, "Yes, I will do a better job of keeping up with the work." This is laudable but it is a rare person who can tell me what practical changes they are going to make. Usually people who do improve can only talk about the changes afterwards, and even then they are sometimes a bit vague. They say things like, "I started working with Hannah every Monday afternoon in the library, and that seemed to help." Being more deliberate would be better, and in this chapter I make lots of suggestions about how to take control of your time, both when planning for a whole semester and when planning for a single week. All of these suggestions can be implemented from day one of your college life, or from whatever day it is when you realize that you're not doing a perfect job of handling the increased freedom and harder work. (If you are planning for revision, you might also want to read Chapter 7; if you are in a state of panic because you are already far behind, start with Chapter 12 then come back to this one.)

One thing to bear in mind, though, is that if you are a normal human being, you will fail to fully follow this advice. You might get distracted by other things, or you might over-shoot and wear yourself out, or you might just find it all a bit intimidating. But it is useful to have something to aim for, and even if you only take up a couple of practical tips, you'll be better off than if you'd never thought about it at all. Certainly it is easier to stay on track, or to get back on track, if you have some idea of where the track is. Before we get into the advice, though, I'll start with a short cautionary section on what you're aiming for and what you're aiming to avoid.

11.2 Aims and things to avoid

First, it's important to have a sense of what qualifies as a particular level of achievement in undergraduate mathematics. You might know this already, but there is some variability in the way that exam scores correspond to grades: to get an A you might need 90% in one course but only 70% in another. That 70% might seem pretty low, but it allows professors to reward students who can reliably do a lot of standard things, while leaving room to give proper credit to those who understand well enough to use their knowledge creatively in novel exam problems.

Second, it's important to understand the implications of retaking undergraduate courses. Most colleges will let you take a course again if you do not get a good grade, and there might be no obvious penalty for doing so. But taking up this option might nevertheless make your life difficult and disrupt your progress toward your major. Your life might become difficult because mathematics is very hierarchical: upper-level courses will rely on material from earlier years. If you don't study well—if you end up knowing only half of the mathematical content—you might still pass these early courses but your chances of doing well in later years will be substantially undermined. Progress toward your major might be disrupted by prerequisite structures: if you choose to retake a course, or if you fail one and are forced to retake it, then keeping on track for your major might mean taking a summer course when you would rather be earning money or working at a summer camp in the mountains or traveling around

Europe. Even when prerequisite rules are lenient, failing to study well in one course can have knock-on effects in others. If you do poorly in Calculus 2, your college *might* let you retake it at the same time as you take Calculus 3. But Calculus 3 will probably rely heavily on material from Calculus 2, and the Calculus 3 professor will assume that you know this material; if you don't, doing the courses simultaneously might be very difficult.

A further, practical reason for wanting to do well early on is that interesting opportunities often depend on it. A certain GPA might be required for getting or keeping a scholarship or a place on an honors program, for instance. If that doesn't concern you, other things might: if you apply for a summer internship, your potential employers will look at how well you have done so far—academically, that's the only thing they have to go on. So, all things considered, it is better and simpler to do well the first time around.

The reason I'm saying this is that, at many colleges, your eventual transcript will record only your best score for each course. If your college is like this then I guarantee that at some point in your studies, some idiot will say to you, "Ah, come on, you can retake the course/get your GPA up later—you don't need to work so hard!" This person has failed to grasp the basic principle that actions have consequences. You don't have to tell them that they're an idiot, but you don't have to let them influence your behavior either.

11.3 Planning for a semester

With all of that in mind, and assuming that you want to do well without making silly amounts of effort, you can do yourself a big favor by doing adequate planning. We'll start by considering large-scale planning and work our way down to the detail.

In planning for a semester, you need to bear in mind that you will have an array of tests and assignments, scheduled at various different times. This might be a new experience if you are accustomed to having a clean cut-off between times at which you learn new material, times at which you revise it, and times at which you are tested. You will also find that there are not many reminders about which work you are supposed to do when. Your professors will expect you to digest the information you are given about tests and assignments, and to work out for yourself how to fit it all in. You might be fine with this, but you might find that your parents and high school teachers supported your time management more than you thought. Either way, the first thing to do is to get a grip on what is happening when, so that you can make sensible choices about when to work on what. This is not hard to do, and you'll get maximum benefit if you do it at the beginning of a new semester. Here is my suggestion for how to proceed.

The first thing to do is to acquire or make a calendar-like wall planner. This might be easy, because such things are often given out by colleges and student organizations. For our purposes, though, mass-produced ones are not ideal because they tend to cover a whole year and have lots of colorful and thus distracting information around the side about events and special offers. For my own planning, I prefer to make one with a simple format.

The next thing to do is to get hold of all the information about when your tests are happening and when your assignments are due. This might take a while if the information is spread across documentation for each individual course, but it is worth the time.

When you have assembled this information, you will find that some tests are scheduled at particular times on particular days. These are easy to add to the planner. You will also find, however, that other things are less well defined. It might be that you have to do an online test, and you can log on and do it at any time during, say, a week. There are several approaches to this. One would be to write it on your planner on the first day it is available. This has the advantage that you'll focus on it early and get it out of the way. You will also feel smug when other people are worrying about it for several more days. Another approach would be to put it on the planner for the penultimate available day. This will give you more preparation time, and might also be better if, say, the earlier days clash with another test. The one thing you absolutely must not do is put it on the last available day. Online testing systems have been around for a while now, and they're pretty robust, but things still do go wrong (occasionally whole college computer networks go down, even). If you leave it until the last minute you will risk causing yourself all sorts of stress. If something bad happens, even if it is not your fault, you will have to chase around trying to get your professor to be lenient with you.

It is also worth considering where on your planner to put other assignments. These will come with a deadline and again, at a bare minimum, I recommend that you pretend the deadline is the previous day. At least once a year someone comes rushing into one of my classes very late, telling a long and detailed story about how the printer didn't work, and then they'd run out of credits, and then they had to go to the other end of campus and all the way back again, and the deadline was 4 o'clock, and then the office was closed.... You get the idea. This person has had a horrible day and, moreover, they have missed other lectures and are now behind with everything else. If you give yourself a spare day, you can't guarantee that problems won't arise, but you can deal with them calmly.

Indeed, the sensible approach might even be to take this to an extreme in some cases. When I was an undergraduate I once had a big mathematics essay that went on for a whole semester and was due after the (four-week) UK Easter vacation. I realized that if I let it run into Easter, it would end up taking the whole vacation, giving me very little time to revise for summer exams. So I worked on it regularly and handed it in before Easter. That was definitely one of the best time management decisions I ever made.

In fact, assignments are varied beasts and it's probably worth thinking beyond straightforward deadlines. If an assignment is short and is given out a week before the deadline, there's not much you can do. If, however, it is a massive thing that is given out in week 3 of the semester and is due at the end, then you have some decisions to make. Most professors don't give out assignments before students have seen at least some of the relevant material, so you should be able to begin straight away. You might decide to look through the assignment instructions, get to grips with what it's asking for, and set some sensible intermediate deadlines (e.g. a date by which you will analyze some data in a certain way). If the assignment involves working in a group, then there are even more things to consider. The problem with other people is that they might be extremely clever and nice, but they might be a bit incompetent or unreliable. In either case it's probably worth scheduling in at least a couple of early meetings in order to sort out what everyone's going to do and when you'll put it all together; in the latter case, you'll value having a pretend deadline or two to allow for things that are out of your control. As always, put these on your planner.

For illustration, suppose you don't have anything that complicated at this stage, and that your planner looks like this:

	Mon	Tue	Wed	Thur	Fri	Sat	Sun
1							
2							
3							
4				German Test			
5	Compu Assign						
6							
7							
8				Calc 4 Midterm			
9	Compu Assign		Lin Alg Midterm	German Midterm			
10	Psych Midterm						
11							
12							
13	Psych Essay						
14	Compu Assign			German Test			

Right away we can see problems coming—week 9 looks like it will be pretty stressful if you don't plan well for it—but we'll come back to that. The next thing to do is to add to the planner any dates on which you will not be around or on which you will not be studying for some reason. For instance, if you are a sportsperson or are heavily involved in a society of some kind, you might have training days or trips that should be blocked out. If you work a 14-hour shift every Saturday, all the Saturdays are out too (or you might be determined that you are not going to study on Saturdays anyway). You will also probably have plans to do things like go home or visit a friend for a weekend. Your planner can help you think about this kind of thing. It might be that you have no control over the dates: if you're going to your grandmother's 80th birthday party, or if your cousin is starring in a Broadway musical and the whole family is going to the opening night, just write it in. If you do have some control, and it turns out that the weekend you were thinking about going away is right before a big assignment is due, you might want to reschedule. There are many things to consider, of course: how well you work at home compared with at college, how often you need a break from your annoying roommate, how long you can stand to be away from your girlfriend, and so on. All I'm advocating is that you make an intelligent choice, rather than an arbitrary one that you'll end up regretting.

At this point, if you are a fan of colorful presentations, you might like to color-code your planner. Personally I am too lazy to walk down the corridor to the color printer so I tend to use different levels of shading. At any rate, what you'll end up with is something like this:

	Mon	Tue	Wed	Thur	Fri	Sat	Sun
1						WORK	
2						WORK	
3						HOME	
4				German Test		WORK	
5	Compu Assign					HOME	HOME
6						WORK	
7						WORK	
8				Calc 4 Midterm		WORK	
9	Compu Assign		Lin Alg Midterm	German Midterm		WORK	
10	Psych Midterm					WORK	
11						AMY'S	AMY'S
12						WORK	
13	Psych Essay					DAD'S B'DAY	
14	Compu Assign			German Test		WORK	

Now you can see all the main things you'll be doing, which is a good place to start.

You might think that what I'm going to suggest next is writing a study plan for the semester. Not a bit of it. In fact, I think that would be a stupid idea. At this point, you don't know how difficult your various courses will be, you don't know when one of your friends is going to suggest that you go apartment-hunting, and you might not have digested any of the schedules for social activities. If you made an upfront plan for the whole semester you'd have virtually no hope of sticking to it and the whole thing would quickly become useless. Instead, at this stage I'm going to recommend planning at the level of a typical week.

11.4 Planning for a typical week

You will hear all sorts of claims about the number of study hours you ought to put in each week. Some people will claim that you ought to be doing two to three hours of independent study for every hour in lectures. This, in my view, is unrealistic. If you have the typical five four-credit-hour courses, it would mean you'd be working something like a 70-hour week. There might be people who can do that, but I'm not one of them. In fact, I wouldn't want to be—if I thought any occupation was going to take up that much of my time and energy, I would seriously consider doing something else instead. Other people (notably your slacker buddies) will claim that you don't need to do anything until it's time to revise for exams. These people are either lying, deluded, or afraid that they're going to fail and determined to take others down with them. Don't be sucked in.

In my view, it is reasonable to aim for roughly a standard working week's worth of hours, say about 40. With about 20 hours already scheduled, that would give you another three hours per week for independent study for each course, plus a small number of spare hours for working on big assignments, preparing for tests, organizing your notes, catching up with anything you've missed, and so on. It ought to allow you to keep up, while also leaving you with time to make the most of college life. Of course, your number might be different. If mathematics brings you such joy that you want to work on it for more hours, go for it. If you are a serious sportsperson, or if you work long hours at an external job, or if you just know that your main aim is to have a great time with your friends, then you might decide that you will be devoting fewer hours to your studies. Whatever you decide, though, once you know how many hours you've got, you can get a realistic sense of what can be achieved in them.

To do that, you need to work out roughly how your hours are going to be distributed in a typical week. I'm often surprised by how little thought students give to this, and I would advocate that there are a few key things you ought to consider. First, you need to arrange your weekly class schedule, and some thought should go into this because, for at least some courses, you will be able to decide which section to attend. To make this decision I think you should ask yourself: At what time of day do I study best? In my experience, few new undergraduate students have thought about this, perhaps because they haven't previously had much choice in how to spend their time. But most of us have some times of day at which we can think well, and others at which we're a bit useless. I, for instance, do my best work first thing in the morning. I'm usually flagging before lunch, and I'm not much good for a couple of hours after it. Then I get a bit more capable again, say 3–7pm. I'm sure any decent dietician would be able to give me instructions on how to even out my energy levels, but I don't have a dietician, and I assume you don't either, so you probably have similar good and bad times.

Your good and bad times might inform your schedule planning in a variety of ways. Perhaps scheduling a few early classes will force you to get up and start the day. On the other hand, perhaps it will just mean that you miss a lot of classes due to over-sleeping. Perhaps you know in advance that one course will be difficult so it would be good to schedule this when you are likely to be alert; perhaps another will be easier so it would be good to schedule this when you *wouldn't* be likely to engage in productive independent study. A schedule will never be perfect, of course, because your choices will be constrained by what is available. But let's say you've considered the options and you've got something like this:

	8:10–9:30	9:50–11:10	11:30–12:50	1:10–2:30	2:50–4:10	4:30–5:50	6:10–7:30
Mon		Linear Alg			Psych		
Tue	Math Compu	Math Compu		Linear Alg	German		Calculus 4
Wed		Linear Alg	Psych				
Thu			German		Psych	Calculus 4	Calculus 4
Fri			German				

Looking at this schedule, we can observe some useful things. Tuesday looks like a bit of a nightmare, for a start. To have any hope of paying attention in that late lecture, you'll have to think carefully about what to do with the open slots. Lectures are demanding if you are properly engaged—the material comes at you thick and fast, and it is hard to stay focused. Because of that, I would probably plan to do no work at all in the open slots; to spend one of them having lunch with friends and the other one at the gym, or something. Friday looks problematic for the opposite reason: you'll be tired by then and, if you don't give it some thought, the lack of scheduled hours could easily mean that you end up doing nothing at all. If you want to study on a day like that, you'll have to think about how to structure your day.

Before we get to that, however, there are some more things that might affect your study patterns. To help you consider these, you might want to make a separate, extended schedule. This one should include the evening hours and weekend days, because it is supposed to give you a genuine sense of how you want to spend your time overall, including the things that you enjoy and any new activities you want to take part in.

The first type of thing to add is regular scheduled activities. Perhaps you have a job in a restaurant, and work two evenings every week. Or perhaps you play hockey and have regular early morning training. Or perhaps your drama group rehearses on a Monday evening and you usually go to the Film Society's showing on a Thursday. Or perhaps your sorority regularly holds parties on a Wednesday. Or perhaps you are a churchgoer so Sunday morning is always busy. You don't want to have to miss these things, and you don't want to end up feeling guilty for doing them, so it's important to factor them in. Whatever you intend to do, block it out on your schedule.

The second type of thing to add is things that will affect whether or not you study because they have a dramatic effect on how alert and capable you are. Perhaps, after your morning training, you are always starving and cannot possibly do anything until you've had a massive breakfast. Perhaps you always drink a lot at parties and are not up to anything until noon the next day. Perhaps you'll be shattered after that long Tuesday in lectures, so there is not really any hope of you doing anything more demanding in the evening than watching trashy TV. Perhaps, in fact, you and your roommates are addicted to a certain evening's TV shows and, although you always tell yourself that you will just watch the first hour and then go to your room and study, this is a big lie and you always end up stuck on the sofa all night. Many of these things you will not be proud of, but you might as well own up to them because, if you don't, they'll happen anyway and you'll be constantly beating yourself up about failing to study when this was never realistic in the first place. If you must, add them to your schedule without shading or in different lettering or something, but do add them. Once you've done that, your full-week schedule will probably look something like this:

	8:10–9:30	9:50–11:10	11:30–12:50	1:10–2:30	2:50–4:10	4:30–5:50	6:10–7:30	7:30–10:30
Mon		Linear Alg			Psych			FILM
Tue	Math Compu	Math Compu		Linear Alg	German	GYM	Calculus 4	TV
Wed		Linear Alg	Psych				WORK	
Thu			German		Psych	Calculus 4	Calculus 4	
Fri			German			GYM		PARTY/OUT
Sat		WORK						
Sun								

Now we're starting to get a more realistic idea of which hours are actually available for independent study, and we can make a rough plan for which ones to use in a typical week. Again, this should involve thinking about when you study well. If you are no use until 10 in the morning, there is no point in trying to study anything complicated before then (though you might be able to do routine exercises or to revise basic material). If you study best in the evenings, you should think about putting in a lot of your hours then. For the sake of illustration, let's say you don't do well in the early morning, you're best in the late afternoon and early evening, and you don't want to study on Saturday. Let's also say you're aiming for around 20 independent-study hours. Then you might block out a rough plan like that shown on the next page. This would give you time to eat lunch every day, a few nights when you can go out without feeling guilty about it, and a lazy Sunday morning. You won't always stick to such a breakdown, but it will serve as a useful framework.

	8:10–9:30	9:50–11:10	11:30–12:50	1:10–2:30	2:50–4:10	4:30–5:50	6:10–7:30	7:30–10:30
Mon		Linear Alg		Study	Psych	Study		FILM
Tue	Math Compu	Math Compu		Linear Alg	German	GYM	Calculus 4	TV
Wed		Linear Alg	Psych		Study		WORK	
Thu		Study	German		Psych	Calculus 4	Calculus 4	
Fri		Study	German		Study	GYM		PARTY/OUT
Sat		WORK						
Sun				Study				

11.5 Planning when to study what

If you've never given much thought to study planning before, then the advice above might be enough for you and you might want to skip over this section. If, on the other hand, you're ready for it, you might go one step further and think about whether it would be good to work on certain courses at certain times. Suppose, for instance, that your Mathematical Computing course is lab-based, and that you can usually finish the work during the lab time. In that case, maybe you only want one independent study hour on that, to make summaries of the main techniques and perhaps get ahead on a coursework assignment. That would mean that for your other subjects, you'd have about five hours each. How should you divide up this time? For mathematics courses, perhaps you want to spend one or two hours reviewing lecture notes (see Chapter 7) and three or four working on problems. I think that's probably reasonable, though obviously it will vary according to the requirements of the course. When would it be sensible to do this work? There is no definitive answer to this, but there are some things to consider. Perhaps, for instance, it's a good idea to review a lecture as soon as possible, so that it is fresh in your mind. On the other hand, perhaps it is good to review one lecture immediately before the next one, so that you can go in prepared. I can't tell you which of these is best (I don't know, and I doubt there's even a meaningful answer), and what you can actually do will be constrained by your schedule. But I do think you ought to consider it, and to think about the trade-offs.

In our example, for instance, all the lectures for Linear Algebra are over by Wednesday morning. So you could, if you wanted to, devote Wednesday afternoon to study of that week's material. This might be great in that it would give you a period of concentration to think about one topic. On the other hand, if it doesn't go well, it might be a disaster—you might spend three of nearly four hours failing to understand notes, failing to do problems, and getting frustrated. You could try to avoid this problem by agreeing that you'll meet a friend in the library and will each spend an hour separately reviewing notes, then half an hour together discussing them, then an hour working separately on problems, then an hour talking through the whole lot. Of course, even if you do that and it works well, you've got another potential problem—it will then be five days before you go to another Linear Algebra lecture, and you might have forgotten it all by then. Maybe your professor will start with a review of the previous week's work, but maybe they won't. So it might be good to move half an hour of your study to the gap before the first lecture on Monday. You see what I mean.

At the other extreme, maybe you'd prefer to spread out your study of any particular course. This has the disadvantage that you wouldn't have a long period of concentrated time, but the advantage that taking a break from something you don't understand sometimes allows you to come to it with fresh eyes. Also, spreading things out might be a good strategy if you absolutely loathe some course. If you loathe something, it's much easier to gear yourself up for an hour of it than it is for four hours of it. Indeed, you might try breaking unlikeable work into even smaller chunks. Say, 20 minutes reviewing a lecture, 20 on a calculation problem, and 20 going over the solutions to a previous problem set. In any case, you could add work on specific subjects to your typical-week schedule, while bearing in mind that things will vary from week to week, especially when tests roll around.

11.6 Planning for an actual week

Recall I said that I think you'd be mad to make detailed study plans a long way in advance. Instead, I suggest a system in which you make a rough plan about once a week. Here is how I would do it.

At the beginning of a week (or, ideally, at the end of the previous one), get out your schedule and your semester planner. Write a list with the name of each of your courses then, for each one, look at the semester planner and write down any specific tasks you need to do that week (coursework, preparing for a test, etc.). Make sure you look ahead when doing this—you might want to start preparing for a big test or begin a coursework assignment that's due in three weeks.

When you've listed these specific tasks, add "lecture notes" and "problems" to each mathematics course, and equivalent routine study tasks to any others. If you've built up a list of questions about a particular course, consider adding "seek help" to that course too. If you're more or less up to date with a particular course, maybe you don't need to add "problems" but you could consider adding "read ahead" instead. So your list might look like this:

Linear Algebra: lecture notes, problems

Calculus 4: test preparation, lecture notes, problems, seek help

Mathematical Computing: start assignment

Psychology: read ahead, essay notes

German: tense exercises, revise vocab

There's a special case to consider at this point, and that's week 1 of a semester. In week 1, you probably won't have any coursework assignments or tests, and you might not have any set problems or reading. In this situation, it's easy to think that you have nothing to do. This is what I hear from advisees sometimes. At the beginning of the semester I'll ask how it's going, and they'll say, "Yeah, fine, we haven't been set any work yet so we haven't got much to do." This response is crazy, on a couple of levels. First, remember that your job at college is not just to wait for set work—you will need to spend time understanding your lecture notes too. Second, even if the lecture notes are straightforward at this stage, there are loads of things you could usefully do. If you're a first-year student, you'll be rusty after the summer and you should get out your high school books and make sure you are fluent with all the material you have studied, filling in gaps where necessary. If you're in later semesters or years, you can review relevant things from previous college courses. Sometimes the names of courses will make this easy—if you are about to start something called "Differential Equations 2," it's probably a good idea to review the key ideas from Differential Equations 1. If the course names don't make it that easy, check which things are prerequisites for your new courses and you'll get the same kind of information. Everyone, new student or not, can also get hold of a course textbook or find out whether any notes are made available in advance, and can read ahead to get a sense of what's coming. You don't need someone to stand in front of you and talk about things before you can study them. So, if it's week 1, all your courses should have the item "review relevant material" and at least some of them should have the item "read ahead."

When you've done that, you've got a list of tasks that could do with some study time this week. In "normal" weeks, this might fit nicely into your typical-week plan, so you can just look at the plan for what subject to work on, look at the list for what to work on in that subject, and get on with it. That's straightforward, and there's a lot to be said for having a routine. Later in the semester, though, when there are more coursework assignments due and more tests to prepare for, the list might start to look a bit alarming. It might have 20 items on it, and you only have 20 hours to play with. Clearly you are going to have to prioritize. Let's think about how.

First, look back at the semester planner, and week 8 in particular. It might be tempting to devote the whole of week 8 to preparation for the Calculus test. But would that be a good idea? It might mean that you're well prepared for the test, but you will also be behind with everything else. In particular, you'll have had a week of not thinking about Linear Algebra and German, both of which have tests coming up less than a week later. Not to mention that assignment for Mathematical Computing. I find that students can be alarmingly short-termist in situations like this. It's as though they are wearing goggles that allow them to see only a couple of days into the future, and the result is that they create a lot of stress for themselves. In a situation like this, some people *will* devote all of week 8 to the Calculus test. Then, when that's over, they'll realize they've done nothing on the Mathematical Computing assignment. They'll spend all weekend panicking about the assignment, but won't be able to finish it because they didn't download something they need, or because they don't understand one of the questions and the professor isn't around to ask and no-one else seems to know either. Then, when they've handed in an incomplete assignment, they'll get wound up because they're a week behind with Linear Algebra and German and they're totally unprepared for the tests. Then, when they've done the tests (not very well), they'll find they're now behind with everything *except* Linear Algebra and German, and they don't understand anything in their lectures any more. They feel out of control all the time. You *can* operate like this as a student, but it's not very pleasant.

I'm not suggesting that you divide up your time as usual in week 8. Obviously it's worth devoting some extra time to studying for the Calculus test. But it probably isn't worth dropping everything else. You want to keep the other things ticking over, at least. One way to do that is to spend the majority of each day's independent study on Calculus but, before starting, to do an hour on something else. This means that you keep the something-else ticking over, but then it's out of the way, which leaves you feeling that you're devoting plenty of time to test preparation. You might also want to add a couple of extra study hours in a week like that, perhaps by missing the film or by booking a shift off work. And, if you do have a week that's atypical, you might like to make an amended actual-week version of the schedule so that you've got something to refer to and you feel in control.

Now, you might find it alarming to think of making all these decisions about what not to do, or what to cut back on for a week or so. I understand that. But, the thing is, if you don't make these decisions consciously, you'll be making them unconsciously anyway. If you avoid thinking about your time, this doesn't give you more hours in a week, it just allows you to be a bit delusional about how well you're using them.

11.7 Where will you study?

Once you've got your schedule, either a typical-week one or an actual-week one, you should give some thought to where you will study. This question might not have occurred to you before. Prior to college, you probably studied in a library or other study room when you were at school, and in your bedroom when you were at home. At college you will have many more options. At my university we have a multi-floored library, two Mathematics Learning Support Centres, general study spaces with computers, general study spaces without computers, various coffee shops and cafés, and probably all sorts of other places that I don't even know about. There's a lot of variety within these places too. Some are big and some are small. Some have big shared desks; some have individual desks with cubicle walls around them. Some have lots of resources such as books and handouts and perhaps people available to help; some are completely bare. Some are used by students working silently on their own; some are used by small groups working quietly together; some are used by noisier groups and have a lot of people passing through as well. Personally, I don't know how anyone gets any work done in the busy places, but they are probably excellent if you have just bought a new outfit and you want to show it off.

What I like, of course, might be different from what you like. You might like working in your room, where you have all your stuff around you, you can put your music on, and you can close the door if you don't want to be disturbed. You might, alternatively, find that all your stuff is a distraction, as is the proximity of the kitchen and your new roommate, and that you'd get more done if you went to a quiet spot in the library. You might develop a liking for a particular floor of the library, but then on some days it might not be practical to go there because all your lectures are on the other side of campus. You see what I mean. The point is that you should think about it, and perhaps try a few places on for size.

11.8 Organizing your stuff

Wherever you plan to study, you will need to make sure that you have the relevant materials with you. This is not rocket science, but it does require a bit of thought because, as a student, you will be the owner of lots of paper, and you will need a system for managing it. If you keep it all in one place, you will end up with a file that is wedged completely full, and probably one or more aching shoulders. You might be okay with that if you like to have everything to hand all the time. What I ended up doing, though, was having a dual filing system. I had a binder for each course, and I kept those in my room. Then I had a separate cart-it-around binder in which I kept just the most recent pages of notes from each course, separated by subject dividers. I usually kept recent problem sets in this binder too.

Regarding separate folders and subject dividers, please do keep your notes and so on properly filed. If you're organized enough to read this book, you're probably organized enough to do this anyway. But I do come across lots of students who have everything stuffed in no order at all into a single flimsy cardboard folder. It's painful to watch these people trying to study. It takes them ages to find what they need, everything ends up dog-eared, and sometimes they lose things altogether. So don't be like that, and don't worry if filing takes a bit of time. It's part of the work, and it takes a lot less time if you do it as things come in than if you try to sort it all out later.

11.9 Not finishing things

Students often believe that once they start something they should keep going until it is completely done. Their high school experience seems to make them think that it's morally wrong on some level to "give up," even temporarily. If you tend to believe that, my advice about planning for fairly short bursts of time might make you think, "But I'll never finish anything!" You're right, of course. Mostly you won't finish things, or at least you won't get to a point where you're thrilled with the result. There will always be a problem set that you never complete, or a bit of your lecture notes you don't understand, or a question on a coursework assignment that you know you didn't answer very well. The thing is, *that would still be true even if you did try to work on everything until you finished it.* When people try to finish things but don't, it's mostly because they can't. They don't get the flash of insight they need, or they get interrupted, or they get tired. Or they end up spending so much time trying to finish one thing that they are forced to do other things in a rush at the last minute.

More importantly, people who try to study like that spend a lot of time not actually studying at all. They might avoid starting something because there isn't enough time to finish it, when in fact starting could be beneficial because it would allow them to mull over the ideas before coming back to the more difficult problems later. Or they might start with the intention of finishing, but get increasingly frustrated and distracted, and spend their time staring out of the window or texting their friends or daydreaming about the girl or boy from Psych 201. Even if the work goes well, there's a law of diminishing returns: you *could* spend another two hours trying to finish a problem set or to get an extra 10 points on an assignment, but would that really be a good use of time? Would you learn very much from it? Could you, right now, make more progress elsewhere and come back to it later? And that's only if you are working well. If you aren't working well— if you're going to spend the two hours sitting there but barely achieving anything—then there's really no contest: let it go, at least for the time being. If you're still uncomfortable about this, read Chapter 13 as well.

It might also be that for you, at the moment, the advice in this chapter sounds good but overwhelming; that you just can't imagine yourself doing everything so systematically. That's okay. You can always try adopting just some of the suggestions, like making a semester planner or thinking about a typical week; you can then add more strategies as you progress through your program.

The last thing to remember about time management is that everyone fails at it, on a regular basis. Even if you make a great start, you will, at some point, get derailed. Assignments turn out to be harder than you think, opportunities come up to do extra shifts at work, people throw each other surprise parties, and sometimes you're in a bad mood and you just don't feel like studying. All of this is perfectly normal, and any of it can throw you off track. The thing is, if you've got a pretty good planning system in place, you can just re-evaluate, make another list, and get back on track quickly. That's what you're aiming for.

SUMMARY

- Most people could improve their time management. Doing so can help a student to keep up with their studies, to feel in control, and thus to enjoy college life.
- Retaking courses will probably be allowed, but there are good reasons to avoid it.
- Mathematics is hierarchical; you should make sure you understand the material in early courses in order to do well in later ones.
- Making a semester planner can help you to get an overview of what is happening when, to plan for weekend activities, and to schedule in coursework and test preparation.
- Planning for a typical week can allow you to think realistically about when you will study. Including social activities on a plan will help to ensure that you can enjoy these without feeling guilty.
- Making a list of current study jobs for different courses can help you to prioritize and decide how to allocate time in a particular week.
- You should give some thought to when and where you study best, and to how to organize your stuff so that you have everything you need with you.
- Students sometimes feel that they have to work on something until they finish it. In reality, this is likely to be ineffective, and I suggest planning for shorter bursts of study time.
- Most people have periodic failures of time management but, if you know what you are aiming for, you should be able to get back on track quickly.

FURTHER READING

For time management tips for students, see:

- Moore, S. & Murphy, S. (2005). *How to be a Student: 100 Great Ideas and Practical Habits for Students Everywhere.* Maidenhead: Open University Press.
- Newport, C. (2007). *How to Become a Straight-A Student: The Unconventional Strategies Real College Students Use to Score High While Studying Less.* New York: Three Rivers Press.

For a practical guide to realistic time management that will take you beyond your studies and into a career, try:

- O'Connell, F. (2009). *Work Less, Achieve More: Great Ideas to Get Your Life Back.* London: Headline Publishing Group.

CHAPTER 12

Panic

This chapter is about why people get behind with their studies, and how to get back on track if you get so far behind that you find yourself in a state of panic.

12.1 Getting behind

Most people get behind at some point during their studies. Some get so far behind that they find themselves in a state of panic, unable to concentrate on any particular piece of work because they're so worried about all the others. If you're reading this chapter because you're in that situation now, turn straight to the "What to do" section and follow the instructions. If you're not, read on.

People get into a state of panic for a number of reasons. Some get into it by slacking off. This might be because they are wilful slackers (fair enough). It might be because they have wilful slacker friends who are always encouraging them to do anything other than study. But it might be because they are ordinary, fallible human beings who let things slide a bit, then find it difficult to get back into the mathematics, then let things slide a bit more, and so on. It's easy to get into a bad situation by this route, and the longer you let it go on, the worse it gets. So be alert to it, especially if you find yourself hiding from your studies by getting unnecessarily involved in other things—taking extra shifts at work when you don't need to, taking on extra responsibilities with a society, obsessively tidying up, things like that. Be especially wary if you find yourself saying "Well, it's nearly the end of semester, I'll catch up when I go home." People who say this often find that it's a lot harder than they anticipated to catch up when they've got incomplete notes and no-one around to support them.

Of course, some people get into a difficult situation through no fault of their own. Perhaps they are ill for a couple of weeks (I mean properly ill, not just a bit sniffley). Perhaps there is some kind of crisis with their housing, or perhaps they suffer a traumatic family event. People to whom this kind of thing happens have every right to feel aggrieved with the world, and to find it doubly hard to concentrate. You shouldn't worry about such things in advance—most people get through their studies without anything worse than a bad cold. And if you do experience something more traumatic, bear in mind that you should be kind to yourself, and that universities have a lot of support services available (see the last section of Chapter 10). In the meantime, the advice in this chapter should help to keep your studies moving.

In between the slackers and the people suffering genuine crises there are people who haven't exactly handled anything badly but who haven't exactly handled it well either. Sometimes, for instance, lots of pieces of coursework are due together; it's easy to end up facing more work than you can actually do in a week full of deadlines. Sometimes people have an early fit of enthusiasm and take on a lot of work for a society or a charity, then find that they can't quite keep up with their studies as well as living up to these promises. People in such a situation should find this chapter useful too, and they should also read Chapter 11.

12.2 What to do

The first thing to do when you're in a state of panic is to be aware that this is actually a normal experience. About once a year someone comes to my office in this state, and that's only among the students I know well—probably there are many more who tell someone else or who handle it on their own. The people who come to me usually seem to relieved when I describe their own experience to them, which I do thus: You feel like you've got a giant ball of work-to-do hanging over your head, and you're walking around looking pretty normal but it follows you everywhere, and every time you try to pull something out of it to work on you find you can't concentrate because you're still worrying about all the other things, so you don't make any progress at all, and you've got to a point where you either can't face any of it or you just find yourself staring blankly at things. The result is that you haven't studied effectively for days or maybe weeks; you can't see how you're ever going to get out of this situation, and you've lost all your confidence.

The good news is that I have a system for turning this around and, almost without exception, people I've used this with come back the next day and tell me they're fine now. I realize that sounds unbelievable but it's actually true. Here is what you do (if you're reading this in a normal state of mind, it will sound weirdly dictatorial and specific but, in a state of panic, specific instruction is what people seem to need):

1. Make a list of each of the courses you are currently studying. Do this on a piece of paper and leave an inch or two of space between each one.

2. For each course, write down the most urgent thing you have to do, together with any associated deadline. This might be a piece of coursework, preparation for a test, or whatever. Look up the deadlines if necessary. If there are two urgent things, list both. If there are no urgent things, just write "notes and problem sheets."

3. Get another piece of paper and write "today" at the top of it (unless you're doing this last thing at night, in which case write "tomorrow" instead). Now pick out the four things from your list that are most urgent overall. Write each one on the "today" paper. If there are more than four things, pick those with the tightest deadlines. If there are fewer than four things, fill the spaces with "notes and problem sheets" for whatever subject(s) you are most worried about. Next to each one, write "30 mins" in brackets.

4. Now consider the first item on the list. You are going to plan how to work on it for 30 minutes. I know that sounds like a very small amount of time, but in your current state you are having difficulty concentrating for a large amount of time, and this will work. What can you do to make progress on that item? Write it down, remembering that you only have 30 minutes so it has to be small. Here are some suggestions:

   (a) If it's a piece of coursework, perhaps you could spend 30 minutes reading the assignment sheet carefully and working out which bits of the notes you'll need to study. Perhaps you've already started it and you need to look something up then try question 2. Perhaps you can't do anything until you have downloaded the appropriate notes or data files, so you need to do that first.

(b) If it's exam or test preparation, and if you haven't done so already, perhaps you could make a summary of the relevant part of your lectures notes. If you have a summary already, perhaps you could try a question from a practice test, or perhaps you could consolidate your knowledge of an important type of calculation by carefully reading an example.

(c) If it's "notes and problem sheets" then, again, perhaps you could make a summary of some section of your notes. If you already have summaries, you could identify an important theorem or definition (one that seems to be used a lot) and try to understand both what it says and how it is used. Or perhaps you could try the first two questions on a problem sheet. That's probably all you can do in 30 minutes.

5. You will now have a list that looks something like this:

TODAY

Linear Algebra test: summarize Chapter 3 (30 mins)

Calculus online test: try easiest questions from practice version (30 mins)

Statistics assignment: download data file and do descriptives (30 mins)

Number Theory: download notes and update definitions list (30 mins)

At this point you might be thinking, "But that is only two hours' work—it's nowhere near enough!" Of course it isn't enough, but you have to start somewhere. That's all we're doing, starting—we'll worry about the rest once we've started. So, if you're feeling a bit flappy, take a couple of deep breaths, get your shoulders away from your ears and calm down.

6. Identify when you're going to do the work. Maybe straightaway, in a block of two hours and ten minutes (the ten minutes is for a break in the middle). Maybe in two one-hour blocks between lectures (though make sure you allow proper time for this—if you have to

walk some distance in a one-hour gap, you can't do an hour's work as well). Identify which piece you'll do when, work out what you will need to have with you to do it, and assemble all the stuff.

7. Pin the list up somewhere you can see it. Or, even better, find a trusted person, show them the list, say "I am going to do these things at these times," and agree a time to check in with them later.

8. When you are ready, make sure you can see a clock, note the time, and start on your first item. Once you've started, if it turns out that you need something that you didn't anticipate, go and get it—you need it, so this is a legitimate use of your time. If it turns out you can't do the question you'd planned without studying some notes, do that studying—again, this is needed for the question, so is a legitimate use of your time.

9. When you have been working for 30 minutes, *stop what you are doing and move on to item 2.* You absolutely must do this, even if it means interrupting yourself. If you don't, you'll get worried about the other items again, and will be back to where you were before following these instructions. Before you start item 2, though, tick item 1 off your list and, underneath it, write down what you need to do next for this course (even if it's just more of the same kind of activity).

10. Do the same for items 2, 3, and 4. If you're doing it all in one go, have a short break in the middle and eat a banana or make a cup of tea or something so that you don't wear yourself out (please don't have loads of chocolate or one of those caffeine-loaded fizzy drinks, though—you want to stay calm, not go off on a stimulant high).

11. When you've done 30 minutes on each of the four items, take a moment to sit back and mentally review what you've done. It's probably quite a lot. In fact, if you're like most people, you'll be surprised by how much progress you've made.

12. Review your list of courses and your "today" list, and make a new list for the next two hours' work.

13. Stop studying for the day if you feel like it (unless you still have lectures or whatever—obviously you should go to those). I really mean this. Two hours' work might not seem like very much but, if you've been in a state of panic, it's probably a lot more than you've done in recent days, and you should feel pleased about it.

14. If you showed someone your list, check in with them. Tell them what you've done, summarize what you've learned, and tell them what you'll do next. This seems to help because it gives you external acknowledgment that you've done what you intended to do. If you think it would be useful, arrange a time to check in with them after your next bits of work too.

In most cases, you will find that following these steps has cured you of your panic. You'll be feeling back in control, and confident that you were suffering from a temporary blip rather than a long-term meltdown. You might still have a lot to do, but you should now feel that you know how to keep making progress (and progress is what you should be aiming for—see Chapter 13 for advice on realistic study expectations and on keeping up). If you still have anxiety because of a serious underlying problem, consider consulting a university support service (see Section 10.9). Otherwise, if you do have another wobble, just run through the steps again.

SUMMARY

- People get behind with their studies due to slacking, illnesses, traumatic events, and just letting things slide.
- If you find yourself in a state of panic, you should be able to get back on track and restore your confidence by listing what you need to do and systematically working on the most important things for short time periods.

FURTHER READING

See the Further Reading section in Chapter 11 for information on how to manage your time and avoid anxiety.

CHAPTER 13

(Not) Being the Best

This chapter is about what mathematical success at the undergraduate level should look and feel like. It aims to dispel some common misunderstandings and replace them with realistic notions about what it means to do well.

13.1 Doing well as a mathematics major

People are usually aware of the practical changes they will experience when they move to college. Some of these are fun and exciting: new friends, much more freedom. Some are mundane: doing your own laundry and living on a tight budget. However, few students seem to be quite ready for the way that the academic experience might feel different. This chapter should reassure you that some experiences might initially seem negative but can be seen as normal consequences of the structure of higher education.

One change that's obvious when you think about it, but that people often don't see coming, is that they are unlikely to be the best in all of their college classes. This is particularly problematic for those who were always the best in their class at high school and who, as a consequence, are accepted by elite institutions. The thing about elite institutions is that they take the best students. At the very top-ranking universities, it is probably the case that almost everyone in the room was the best in their class at high school. At other universities and colleges, it might be far fewer people, but there will still be a high concentration of mathematical success and talent.

This means that many people experience a change in their intellectual status at some point during their college life.

People who were not the best in their high school tend to cope with this without major problems. They sometimes suffer from worries, of course—they might think that a mathematics major is only for the best of the best, and they might be concerned that they won't live up to expectations. But usually they have developed some skills for coping with such worries, and they are generally not disturbed by temporary difficulties; they don't tend to interpret these as signs of impending long-term failure.

People who have been the best in their class at high school, however, sometimes suffer a crisis of confidence, because they tend not to realize how much of their sense of identity is tied up with being good at mathematics. In our culture, many people think that mathematical ability is particularly indicative of intelligence, and that it is based on inherent talent rather than hard work. This view is questionable but very pervasive. As a result, those who have always been good at mathematics tend to be thought somewhat special, and this gets built into their sense of why they are worthwhile human beings. So, if they no longer stand out, it can shake their sense of their place in the world.

In fact, at college, I do not think that "the best" is meaningful; there will not be one person who is "the best" in any given intake of students. Undergraduate mathematics is more varied than you might think. Different people will like different things and excel at different things, and someone who studies in an organized way might easily do better than someone who is "brighter" on some measure but who slacks off. It's not clear that those who stood out in high school will stand out during their major, and it's not clear that excelling at undergraduate mathematics leads naturally to being successful in postgraduate study or mathematical careers. These things are probably correlated, but they also demand somewhat different skills. Everyone should find this encouraging—you might discover that you have a real flair for something you've never even considered before.

That said, experience of doing well in high school mathematics does seem to set up expectations that are problematic if applied directly to undergraduate study. So here is some information about common expectations and about how to be realistic.

13.2 What does understanding look like?

Periodically, I meet a student who is worried because they used to think they were good at mathematics and now they think they're not. When I ask about this in more detail, it usually turns out that they have an inaccurate idea of what being good at mathematics should look and feel like. In particular, they often have unrealistic expectations about how their understanding of undergraduate mathematics should develop.

Often, it turns out that this person used to understand just about everything their high school teachers said, but they don't understand anywhere near everything their professors say. They take this to mean that they don't understand undergraduate mathematics. But there is an error in this reasoning. They're essentially equating "mathematics I can understand" with "mathematics I can understand straightaway," when in fact the first of these two sets is probably a lot bigger. I'm sympathetic, because someone in this position has probably experienced a dramatic shift in how they view themselves in relation to their studies. If you understand most of what is said, you will feel good about it, and you probably won't have to do much work to get to a point where you understand it all. If, suddenly, you don't understand very much, you will feel bad about it, and you will probably have to do a great deal of work to get to a point where you understand it all. Obviously that's daunting.

If you find yourself in this position, the first thing to do is to recognize that it's entirely normal. The fact that you don't understand everything in lectures doesn't mean that you are not good at mathematics any more, it just means that the mathematics is harder and the pace is faster. To keep up with it, you will have to do more (or better) work. But so will everyone else. And that's as it should be (if studying for a mathematics major was easy, everyone would do it). Once you've recognized this, there's really nothing for it but to stop devoting your energy to worrying, and start making a dent in the work. Often this won't be as hard as you think, especially if you approach the task with good strategies (see Chapters 7 and 11) and with realism about what you should be trying to achieve (see below).

13.3 Keeping up

Some people get hung up on the fact that they don't understand something at the beginning of a course. They think, "If I can't even understand the beginning, how will I understand the rest?" This can lead them to spend a lot of time worrying while the course moves on ahead of them. But this thinking, too, contains an important error in reasoning. It might be that this beginning item (call it item A) is very important for the rest of the course; that the course is structured like this:

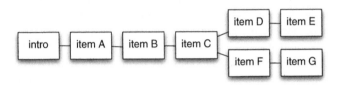

But it might be that item A is needed for just one method, or is part of an interesting theoretical aside but is not key to the rest of the course; that the structure is more like this:

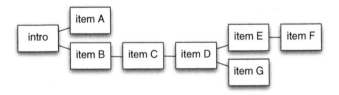

Therefore, before devoting lots of energy to worrying about item A, it's a good idea to have a stab at working out how important it is. If lecture notes are provided in advance and the concept seems to come up all the time, it is probably worth sorting it out. If not, it might be something you can safely put aside and come back to later, and your current effort might be better spent on something that's central for the whole course, or on something that's going to be used a lot in the next couple of weeks.

A related problem is faced by students who get behind in a course. Being behind presents you with a choice: do you try to catch up by going back to where you last understood it and working forwards, or by trying to understand the new material as it appears, and going back to fill in the gaps according to what is currently needed? I would favor the latter option, for the following reasons.

First, the working-forwards option is emotionally unpleasant because you will probably always be behind. It is also practically wasteful because it means that you'll spend a lot of time sitting in lectures you don't understand. Second, the start-from-here-and-go-back-to-fill-gap option fits in better with the strategies for learning from lecture notes discussed in Chapter 7. Making it work requires a bit of thought, though, and I suggest the following. If the course has lecture notes that are given out in advance, look at what's coming up in the next lecture. If it doesn't, look at the notes from the most recent lecture instead. Either way, identify any key concepts or theorems that are used, and go back to study those: look up the relevant definitions, examples, and proofs, study them using the strategies from Part 1, and try any associated problems. That way, you'll fill some gaps, and you'll be better placed to follow the next lecture. Do this a few times and you'll probably find that you feel you have caught up, even though you still have some gaps. Indeed, you should develop a pretty good grasp of the structure of the course, because you'll be focusing on links between earlier and later material. You would be unlikely to reach this point if you just sat in lectures copying things down. This is a good example of a situation in which taking charge of your own learning is key.

13.4 Understanding and speed

Lots of people think that being good at mathematics means being fast at it. This belief seems to be associated with several errors in reasoning, so let's unpick the logic and work out how the two actually fit together.

It is certainly the case that, all else being equal, being fast at some piece of mathematics is better than being slow. Being fast means that you'll have more time to do more practice, or to study another topic, or to go out with your friends. It is also the case that people who are good at some mathematical topic—those with a deep and thorough understanding of it—are often fast at solving associated problems, whether these involve routine calculations or complex proofs. However, the fact that people who are good are often also fast does not tell us how they got to be fast.

They might have been fast from day one, or they might have spent a long time carefully studying details, and only sped up when they'd mastered things properly. Trying to mimic them by being fast from the outset is not necessarily the best way to be fast in the end.

The problem with trying to be fast for its own sake is that it is not very compatible with developing the deep and thorough understanding that is necessary for long-term mathematical growth. People who try to be fast often miss out on opportunities to consolidate their understanding, because racing on to the next problem as soon as they finish one means that they do not reflect on what they have learned, do not think about how it relates to things they already know, and do not tighten their grip on it so that they'll remember it and be able to apply it in new situations. Reflecting might only take a minute, and it might dramatically increase your understanding of what you have done (Section 1.2 suggests some reflection questions).

Also, people who try to be fast often make mistakes that they could have avoided. Sometimes they do this in a bid to impress others—to demonstrate quick and insightful thinking to their teachers and fellow students. I recently had a student, for instance, who was almost always first to answer when I asked a question. His first answer was often wrong. He would then give a second answer that was correct and that showed good insight. It wasn't that he couldn't do the thinking, it was that he didn't give himself time to do it properly. I'd have been more impressed if he had waited a minute, thought properly, and then given the better answer as a first attempt.

In the worst case, trying to be fast can lead people to memorize procedures without understanding what they are doing at all. As discussed in Chapter 1, this might occasionally be an appropriate way to proceed, but it's worth thinking about whether it is really helping you achieve what you want to achieve in the longer term.

13.5 Not trying to understand everything

Returning to the subject of realism, it's important to bear in mind that you probably won't understand everything in a given course, even by the time of the exam. That might sound horrifying, especially if you are a dedicated student. But it's a natural consequence of the change in the mathematical environment. A mathematics major is supposed to challenge even the brightest and most hardworking students. When the class is full of such students, it has to be very challenging indeed.

This isn't as bad as it might sound, especially if you recall that you do not necessarily need to know everything to get an A (see Section 11.2). One consequence of this is that you are free to stop worrying about some things. By the time of the exam, there will be methods that you can't reliably apply and proofs that you can't get your head around. But that is okay. Indeed, it might be a good idea to *aim* for such a situation. That might sound horrifying too, so I will explain. There are two basic reasons. First, in many courses, it is better to understand a large chunk of the material well than it is to understand all of it badly. Imagine that a student has tried to master 100% of the material, but does not know it very well, so they can only reliably get 60% on any given question. That gives an expected score of 60% overall. Now imagine that they have only mastered 80% of the material, but know it well enough to be confident of getting 90% on any question on that material. 90% of 80% is 72%. That's much better. My experience is that some students take to this notion naturally, but others find it very hard to accept. They seem to feel that they are cheating the system by not trying to learn absolutely everything. I'm not sure what to say about this, except that it indicates a misunderstanding of what the system would like you to achieve. Mathematicians don't want students to end up with encyclopedic knowledge but dodgy underlying understanding. They're more impressed by deep understanding, and, at least in upper-level courses, they tend to test for this by making at least some of the questions on their exams require original thinking or novel applications of the mathematical ideas. Doing well thus requires good problem solving skills and a solid understanding of the concepts and methods from the course as well as the relationships between them. You can't prepare for that by memorizing ever more stuff.

The other reason that you might consider not learning some of the material is that devoting the same amount of energy to every part of the course is unlikely to be effective, for the following reason. In any given course, there will be some things that you understand well, some things that you kind of get but are not really sure about, and other things that you find totally mystifying (if you have read Chapter 7, you will recognize this argument from Section 7.5 on using summaries for revision). If you do the same amount of work on each kind of thing, you will waste some time going over things you don't really need to study, and more time bashing your head against a brick wall with the totally mystifying things. I would argue that you'd be better off working primarily on the things that you are not really sure about. With these things, you are likely to make some progress. This will feel good. Also, while making progress, you are likely to revise the easier things, because you will have to use them in new ways. This will involve making new connections among the easier things, which is arguably better than just practising them over and over again. Finally, the progress will mean that some of the previously mystifying things will now seem more accessible. You probably won't sort out all of the mystifying things but, as argued above, that's fine.

Obviously understanding everything is a fine aim, and you should certainly begin every course with the intention of grasping all the main ideas and principles. You probably won't manage it every time, but that's okay. What you shouldn't do is take this as evidence that you're not very good any more, or as an indication that you should give up. Take it as an occasion to rise to the challenge, instead.

13.6 The mythical genius

At this point it's probably worth saying something about the mythical genius who does no work at all but who understands everything immediately and gets an A in every course. You will doubtless hear tales of such students—people like the romantic idea of the genius who can do everything effortlessly. I do not cast doubt on the existence of such students, but I do think they're much fewer and further between than a lot of people would have you believe. If you have heard about one, admire them by all means, but not before you've asked yourself two questions.

First, is this person actually doing as well as everyone seems to believe? I say this because I was once told by a couple of students that their friend was one of these people—that he just picked everything up with no effort. Incredible as it sounds, I later found out not only that he was not doing outstandingly well (B grades, no better), but also that he had actually *failed* many of his courses the previous year and was repeating them! You have to admire this student—he'd done a cracking job of pulling the wool over his friends' eyes. Or, more charitably, perhaps his friends had acted impressed one day and he hadn't had the heart to tell them the truth. In any case, stories you hear about people do not necessarily reflect their actual abilities. Be aware of this before you get too starry-eyed.

The second question to ask is, does the person really do no work? Or is it possible that they actually do quite a bit of work, but do it on the quiet? Some people are quite shy and like to get on with things without making a big fuss. Others like to cultivate a devil-may-care image, but are prepared to work hard in the background to make sure that they keep on top of things. Either way, just because this person is not often *seen* to be studying, that doesn't mean that they're not. In fact, I do not think it credible that a really talented person would do next to nothing. People who are good at things tend to like doing them. I guess it's possible that they'd spend no time studying for their courses because they'd found an interesting book on an advanced topic and were studying that instead, but even then they'd be practising the types of reasoning that would be useful in undergraduate work.

All of this said, you might well run across an extremely talented person during your studies. At the very least, you will likely become friends with someone who is substantially better than you at some subjects. You will probably be a bit in awe of this person, and you might quite enjoy the experience, because being in awe of someone is emotionally engaging— a bit like being a fan of a really great band/skateboarder/TV sci-fi writer, except that you actually know the person and can show off about it to your other friends. But, if you find yourself in this situation, I think you should ask yourself one more question, which is really the important one:

What can I learn from this person?

The reason you should ask this is that no matter how much talent this person has, the mathematics doesn't just jump into their brain unbidden. Their talent must manifest itself in actual thoughts and actions and, if you can find out what some of these are, you can probably get better at the subject yourself. If you are lucky, your friend might be able to tell you directly what they are doing to develop such a good understanding so quickly. Perhaps they always think of a certain concept by relating it to a certain diagram, which makes memory easy and relationships obvious. Perhaps they always start their study by reading ahead to identify the main theorems in the course, then relating material to those as they go along. There are all sorts of things they might be doing that might not have occurred to you, but that you might be able to do pretty well once you've heard about them. If you're less lucky, your friend might not be able to tell you what they are doing, because it comes naturally to them so they've never reflected on it. Even so, by engaging them in conversation about how to approach various types of problem, you might be able to infer some useful ways of thinking.

Of course, you have to be thoughtful about this, as you always do when working with others. Mimicking someone's outward behavior is not the same as doing the thinking—you have to do more than just mindlessly copy. But, as discussed in Chapter 10, human beings don't exist in isolation, and a great amount of what you learn in life you will learn from other people. If you are willing to share your knowledge and understanding, and to do so in a way that is friendly, equitable, and allows everyone to participate, you will find that others are willing to return the favor.

SUMMARY

- Many people experience a change in their intellectual status as they progress through a mathematics major. Some suffer a temporary crisis of confidence because of this.
- Undergraduate mathematics is more difficult than high school mathematics and is presented at a faster pace; finding this a challenge is normal, and does not mean that you are not good enough any more.
- Keeping up might be easier if you focus your attention on the central ideas in a course or on those which are likely to be used in upcoming lectures.
- Being good at mathematics is not necessarily the same as being fast at it. Taking the time to understand things properly is worthwhile.
- It is normal not to understand everything by the time of an exam; knowing this, you should think about how to distribute your effort in order to do well.
- You will probably hear myths about genius students, and you will certainly find that others sometimes understand things better than you do. If you are willing to share your knowledge, others will reciprocate.

FURTHER READING

For more on realistic expectations and adjusting to university life, try:

- Moore, S. & Murphy, S. (2005). *How to be a Student: 100 Great Ideas and Practical Habits for Students Everywhere*. Maidenhead: Open University Press.
- Newport, C. (2007). *How to Become a Straight-A Student: The Unconventional Strategies Real College Students Use to Score High While Studying Less*. New York: Three Rivers Press.

For more on reflecting on your knowledge and becoming a better problem solver, try:

- Mason, J., Burton, L., & Stacey, K. (2010). *Thinking Mathematically (2nd Edition)*. Harlow: Pearson Education.
- Pólya, G. (1957). *How to Solve It: A New Aspect of Mathematical Method (2nd Edition)*. Princeton, NJ: Princeton University Press.

CHAPTER 14

What Mathematics Professors Do

This chapter describes what mathematicians do when they are not giving lectures. It aims to give an idea of what it is like to be a professor, and of the way that undergraduate education fits into a university's many activities.

14.1 When professors aren't lecturing

Often, undergraduate students do not have a very clear idea of what a professor's job involves. You can be a successful undergraduate without one, of course, but personally I think that it's interesting to learn about the wider activities of people you come into contact with every day. Also, you never know, you might want to become a mathematician one day yourself.

The fact that students don't know what professors do is not surprising given their experience in high school. If a high school teacher is not teaching you, they are usually teaching somebody else, or preparing to teach, or grading students' work, or going to training sessions to learn new teaching techniques. More senior teachers also spend time managing the school: deciding how to spend the budget, hiring new staff and overseeing their training, liaising with the school board, attending educational meetings, and so on.

For university professors, the job is rather different. Across the year a typical professor might spend only about 40% of their time on teaching-related activity; they spend perhaps 10% doing administrative work, and the remaining 50% doing research. Here is a bit of information about each type of work.

14.2 Teaching

Some of the teaching that professors do is visible to everyone. For instance, I give lectures (obviously), I run recitations, I see a small group of first-year advisees for an hour a week, and I spend two hours a week in our drop-in Mathematics Learning Support Centre. I also write lecture notes, problem sheets and solutions, post these on the VLE, grade my advisees' coursework, liaise with the PhD students who act as teaching assistants, and respond to students' emails. I write exams and resits and adjust them according to feedback from an internal moderator (another professor in my department) and an external examiner (an experienced professor from another university who vets all the department's exams to make sure that standards are being maintained). Finally, I mark all the exams for my courses. This takes forever and is inordinately dull—if you see a professor looking glassy-eyed with boredom shortly after the exam period, you will know why.

Other teaching, however, is less visible because it involves fewer students. For instance, I usually have one or two final-year project students (see Chapter 10), each of whom I see for about an hour a week. I have a student on an internship (again see Chapter 10); twice a year I go to visit them at their company, and I grade two reports they write on their work. I also keep up with my second- and third-year advisees, seeing each of them individually a couple of times a year. I enjoy this part of my job—it's nice to watch them become more experienced and start making decisions about what to do with their lives. I work in the UK now so your professors might have different teaching responsibilities, but they probably have an array of work somewhat like this.

14.3 Administration

"Admin" refers to things that keep the department running smoothly. Some of this is done by administrators, who maintain student records, collect and return coursework, collate coursework and exam scores, arrange open days, and so on. Some of it, though, is done by academics (professors often refer to themselves in this way). For instance, professors answer queries from potential students, give open day talks, assign teaching tasks, act as mentors to new professors, arrange for external speakers to give departmental seminars, update information about various degree programs, run the department's website, assess promotion applications, and attend institutional meetings about various aspects of finance and development strategy.

Most professors are involved in some way in activity like this. Those who are fairly junior usually have a small role that doesn't take up much of their time. Those who are more senior might have a large role with a lot of responsibility. Indeed, someone will be Head of Department—a big role that involves running the department and representing it at higher-level committees. Sometimes people with big jobs like this are not expected to do any teaching at all, so don't be surprised if your Head of Department doesn't actually give lectures.

14.4 Research

Many professors spend a large amount of time doing research, which in the case of mathematicians means developing new mathematics. This sometimes surprises new undergraduates who, if they've thought about it at all, tend to assume that mathematics is already "finished." But, as I said in Chapter 4, the quadratic formula has only been known for about 400 years; much of what you'll learn as an undergraduate was developed about 200 years ago and, depending on your department's staff and your option choices, you might be learning mathematics that has only been around for 20-odd years. New mathematics is developed all the time by the thousands of research-active mathematicians around the world.

The aim of mathematical research depends upon a mathematician's specialism and research interests. Pure mathematicians work with abstract structures, proving general results about relationships across these structures. You can probably get a sense of what this feels like if you think about the discussion in Chapter 2, which involved thinking about abstract objects and their properties. Pure mathematicians usually work on these structures for their own sake, but the results often turn out to be useful in real-world applications. Applied mathematicians work more directly on mathematics with an obvious application, perhaps designing and testing increasingly accurate models of real-world phenomena like brain activity or the spread of diseases in a population. There is middle ground, however, and it is common for mathematicians to collaborate with others from different specialisms or disciplines.

As such, the form taken by mathematical research involves quite different types of activity. Some mathematicians work with pen and paper, filling pages with handwritten calculations and attempted proofs. Others make use of substantial computing power to perform complex calculations or to run simulations.[1] All spend a lot of time reading, keeping up to date with papers in research journals in order to learn about progress made by others in their field. Perhaps someone has proved a useful new theorem, or produced a new, more elegant proof of an existing theorem, or shown that a theorem can be generalized in an interesting way. Perhaps, in fact, some mathematicians are less interested in the theorem than in the proof; they might read in order to adapt the ideas to their own current research problem. Then, of course, mathematicians attempt to get their own work published in these same journals. This involves submitting a paper on the work, which will then be sent out to a small number of other mathematicians for review. The reviewers might request changes if they think that there is an error or an inadequate explanation, and the author(s) will then amend the paper, if they can, before the journal editor eventually decides it is worthy of publication (or rejects it as inadequate). So mathematicians are also involved in this type of review process, and some of them work as journal editors.

Mathematicians are also involved in a lot of spoken discussion. Many work with PhD students, setting them problems to work on, pointing them toward what to read, and supporting them as they become less like undergraduate learners and more like independent researchers. Mathematicians also discuss their work with each other, sometimes informally—a mathematician might realize they need a particular concept for their work and decide to speak to a colleague who is more knowledgeable about it—and sometimes formally, taking part in local, national, and international seminars, workshops, and conferences. These events are somewhat different in character (as explained below) but they usually involve people giving talks

---

[1] Notice that it's not the computer doing the intellectual work in this second case—the mathematician has to decide what's worth calculating, program the computer to do it, and interpret and report the results.

while others listen. This is a bit like being in a class except that everyone is more like a teacher than a student; people ask challenging questions, demand justifications if they are not sure about a particular argument, get into debates about the finer points of a theory, and refine and extend their own understanding by working to see how it relates to the claims made by the speaker and by others in the room.

Seminars, usually, are fairly small and involve just one person speaking. Your department will probably have a professor who acts as seminar organizer, inviting mathematicians from other universities to speak on topics that are of interest to particular research groups; keep your eyes open and you will probably see lists of upcoming seminars on noticeboards or information screens. Workshops are usually somewhat larger. Some are regular meetings involving an established research group, and some are one-off events where a wider group of people with similar mathematical interests spend a few days presenting work to each other and discussing its merits and consequences. Conferences are usually the biggest; the largest international conferences bring together as many as a few thousand mathematicians. These people don't all have the same interests, so big conferences usually have a combination of sessions of various sizes. There might be keynote talks given by internationally famous mathematicians and attended by hundreds of people, and smaller sessions on more specialist topics, 20 or more of which happen simultaneously so that people can decide what they want to attend. Some conferences are one-day events, but others run for several days and involve everyone in drinking a lot of coffee during the day and, depending on their proclivities, a lot of alcohol in the evening. As in all academic subjects, the social aspect of such events is considered very important—it is often in informal conversation that people catch up with their colleagues' new ideas and develop collaborations. Because of this sort of meeting, most people who work in an academic field for any period of time develop close friendships and working relationships with people from all over the country and the world. Sometimes these collaborations lead to further travel; it is not uncommon for a mathematician to take a sabbatical and spend a few months working with colleagues at another institution.

The upshot of all this is that there is a lot going on at a university, and that the job of professor is interesting and varied. It's not a career that's right for everyone, but I highly recommend it if you are self-motivated and interested in developing knowledge for its own sake. The money's not bad and the freedom to work largely when you want, where you want, and on what you want, is extraordinary.

14.5 Becoming a mathematician

I started this book by saying that I hoped its content would be useful to anyone who wants to study as a mathematics major or who is already doing so. I hope it has. I hope it has given you some insights that will help you to focus on the right things and to enjoy your studies. I also hope that you will continue to find it useful as you progress through your major; if you return to it periodically, you will probably find that you are developing a more nuanced view of the ideas the book covers.

I certainly hope that you will be successful in your studies, and that you will complete them with a feeling that you have developed a deeper appreciation of the power and beauty of mathematics. I hope, in fact, that you come to see that the habits of mind you have developed will be useful in any walk of life, whether or not you end up using mathematics in your career. Some readers of this book, for instance, will go into general graduate jobs as retail managers or human resources people or management consultants or international sales executives. They will probably never differentiate a function again, but they will regularly use the skills they've developed in constructing logically sound arguments, in critiquing arguments made by others, and in presenting their conclusions clearly in both written and spoken communication. Others will go into more obviously mathematical careers, as accountants or actuaries or statisticians or financiers or teachers. These people might not use all of the mathematics they've learned in their degree, but they will find that some of it is put to work in new applications, and that their skills in mathematical problem solving are useful on a daily basis.

I do hope, however, that some readers do want to become mathematicians. I hope that they'll enjoy a particular course, begin to specialize in that area, do a project to experience more independent and original work, study for a PhD under an inspiring supervisor and go on to be a successful research mathematician. Others have more to say than I do on the next stages of this process, so I refer you to the further reading section if this applies to you.

Whatever your individual situation is, I'm pleased that you want to study mathematics because it's a fantastic subject, and I wish you the very best.

SUMMARY

- Mathematics professors are usually involved in teaching, administration, and research activities. Some of this will be visible to an undergraduate student, but some goes on in the background.
- Mathematicians who are active in research usually work with PhD students, publish research papers, and participate in local, national, and international seminars, workshops, and conferences.

FURTHER READING

To see a list of mathematics journals (and to explore the contents of some, depending on access permissions at your university), click on the link to Mathematics at:

- http://www.jstor.org/

For more information on career options for mathematics majors, try:

- http://www.maa.org/careers/
- http://www.ams.org/careers/
- http://www.siam.org/careers/thinking.php

For more information on becoming a mathematician, try:

- Stewart, I. (2006). *Letters to a Young Mathematician*. New York: Basic Books.
- Terence Tao's website at http://terrytao.wordpress.com/career-advice/

BIBLIOGRAPHY

Aberdein, A. (2005). The uses of argument in mathematics. *Argumentation*, *19*, 287–301.

Ainsworth, S. (2008). The educational value of multiple-representations when learning complex scientific concepts. In J. K. Gilbert, M. Reiner, & M. Nakhleh (Eds.), *Visualization: Theory and Practice in Science Education* (pp. 191–208). New York: Springer.

Alcock, L. (2010). Mathematicians' perspectives on the teaching and learning of proof. In F. Hitt, D. Holton, & P. W. Thompson (Eds.), *Research in Collegiate Mathematics Education VII* (pp. 63–92). Washington DC: MAA.

Alcock, L. & Inglis, M. (2008). Doctoral students' use of examples in evaluating and proving conjectures. *Educational Studies in Mathematics*, *69*, 111–129.

Alcock, L. & Inglis, M. (2010). Representation systems and undergraduate proof production: A comment on Weber. *Journal of Mathematical Behavior*, *28*, 209–211.

Alcock, L. & Simpson, A. (2001). The Warwick analysis project: Practice and theory. In D. Holton (Ed.), *The Teaching and Learning of Mathematics at the Undergraduate Level* (pp. 99–112). Dordrecht: Kluwer.

Alcock, L. & Simpson, A. (2002). Definitions: dealing with categories mathematically. *For the Learning of Mathematics*, *22*(2), 28–34.

Alcock, L. & Simpson, A. (2004). Convergence of sequences and series: Interactions between visual reasoning and the learner's beliefs about their own role. *Educational Studies in Mathematics*, *57*, 1–32.

Alcock, L. & Simpson, A. (2005). Convergence of sequences and series 2: Interactions between nonvisual reasoning and the learner's beliefs about their own role. *Educational Studies in Mathematics*, *58*, 77–100.

Alcock, L. & Simpson, A. (2009). The role of definitions in example classification. In M. Tzekaki, M. Kaldrimidou, & H. Sakonidis (Eds.), *Proceedings of the 33rd International Conference on the Psychology of Mathematics Education* (Vol. 2, pp. 33–40). Thessaloniki, Greece: IGPME.

Alcock, L. & Simpson, A. (2011). Classification and concept consistency. *Canadian Journal of Science, Mathematics and Technology Education, 11*, 91–106.

Alcock, L. & Weber, K. (2010). Referential and syntactic approaches to proving: Case studies from a transition-to-proof course. In F. Hitt, D. Holton, & P. W. Thompson (Eds.), *Research in Collegiate Mathematics Education VII* (pp. 93–114). Washington, DC: MAA.

Alibert, D. & Thomas, M. (1991). Research on mathematical proof. In D. O. Tall (Ed.), *Advanced Mathematical Thinking* (pp. 215–230). Dordrecht: Kluwer.

Allenby, R. B. J. T. (1997). *Numbers & Proofs.* Oxford: Butterworth Heinemann.

Almeida, D. (1995). Mathematics undergraduates' perceptions of proof. *Teaching Mathematics and its Applications, 14*, 171–177.

Antonini, S. (2011). Generating examples: Focus on processes. *ZDM: The International Journal on Mathematics Education, 43*, 205–217.

Arcavi, A. (2003). The role of visual representations in the learning of mathematics. *Educational Studies in Mathematics, 52*, 215–241.

Asiala, M., Brown, A., DeVries, D., Dubinsky, E., Matthews, D., & Thomas, K. (1996). A framework for research and curriculum development in undergraduate mathematics education. In *Research in Collegiate Mathematics Education II* (pp. 1–32). Washington, DC: American Mathematical Society.

Asiala, M., Dubinsky, E., Matthews, D. W., Morics, S., & Oktac, A. (1997). Development of students' understanding of cosets, normality, and quotient groups. *Journal of Mathematical Behavior, 16*, 241–309.

Bardelle, C. & Ferrari, P. L. (2011). Definitions and examples in elementary calculus: the case of monotonicity of functions. *ZDM: The International Journal on Mathematics Education, 43*, 233–246.

Bell, A. W. (1976). A study of pupils' proof conceptions in mathematical situations. *Educational Studies in Mathematics, 7*, 23–40.

Bergqvist, E. (2007). Types of reasoning required in university exams in mathematics. *Journal of Mathematical Behavior, 26*, 348–370.

Biggs, J. & Tang, C. (2007). *Teaching for Quality Learning at University.* Maidenhead: Open University Press.

Biza, I. & Zachariades, T. (2010). First year mathematics undergraduates' settled images of tangent line. *Journal of Mathematical Behavior, 29*, 218–229.

Brown, J. R. (1999). *Philosophy of Mathematics: An Introduction to the World of Proofs and Pictures*. New York: Routledge.

Buchbinder, O. & Zaslavsky, O. (2011). Is this a coincidence? The role of examples in fostering a need for proof. *ZDM: The International Journal on Mathematics Education, 43*, 269–281.

Burn, R. P. (1992). *Numbers and Functions: Steps into Analysis*. Cambridge: Cambridge University Press.

Burn, R. P. & Wood, N. G. (1995). Teaching and learning mathematics in higher education. *Teaching Mathematics and its Applications, 14*, 28–33.

Burton, L. (2004). *Mathematicians as Enquirers: Learning about Learning Mathematics*. Dordrecht: Kluwer.

Chater, N., Heit, E., & Oaksford, M. (2005). Reasoning. In K. Lamberts & R. Goldstone (Eds.), *Handbook of Cognition* (pp. 297–320). London: Sage.

Chi, M. T. H., Bassok, M., Lewis, M. W., Reimann, P., & Glaser, R. (1989). Self-explanations: How students study and use examples in learning to solve problems. *Cognitive Science, 13*, 145–182.

Chi, M. T. H., Leeuw, N. D., Chiu, M.-H., & LaVancher, C. (1994). Eliciting self-explanations improves understanding. *Cognitive Science, 18*, 439–477.

Coe, R. & Ruthven, K. (1994). Proof practices and constructs of advanced mathematical students. *British Educational Research Journal, 20*, 41–53.

Conradie, J. & Frith, J. (2000). Comprehension tests in mathematics. *Educational Studies in Mathematics, 42*, 225–235.

Copes, L. (1982). The Perry development scheme: A metaphor for learning and teaching mathematics. *For the Learning of Mathematics, 3*(1), 38–44.

Cornu, B. (1991). Limits. In D. O. Tall (Ed.), *Advanced Mathematical Thinking* (pp. 153–166). Dordrecht: Kluwer.

Crawford, K., Gordon, S., Nicholas, J., & Prosser, M. (1994). Conceptions of mathematics and how it is learned: The perspectives of students entering university. *Learning and Instruction, 4*, 331–345.

Crawford, K., Gordon, S., Nicholas, J., & Prosser, M. (1998a). Qualitatively different experiences of learning mathematics at university. *Learning and Instruction, 8*, 455–468.

Crawford, K., Gordon, S., Nicholas, J., & Prosser, M. (1998b). University mathematics students' conceptions of mathematics. *Studies in Higher Education, 23*, 87–94.

Credé, M., Roch, S. G., & Kieszczynka, U. M. (2010). Class attendance in college: A meta-analytic review of the relationship of class attendance with grades and student characteristics. *Review of Educational Research, 80*, 272–295.

Dahlberg, R. P. & Housman, D. L. (1997). Facilitating learning events through example generation. *Educational Studies in Mathematics, 33*, 283–299.

Davis, P. & Hersh, R. (1983). *The Mathematical Experience*. Harmondsworth: Penguin.

de Jong, T. (2010). Cognitive load theory, educational research, and instructional design: Some food for thought. *Instructional Science, 38*, 105–134.

de Villiers, M. (1990). The role and function of proof in mathematics. *Pythagoras, 24*, 17–24.

Deloustal-Jorrand, V. (2002). Implication and mathematical reasoning. In A. D. Cockburn & E. Nardi (Eds.), *Proceedings of the 26th International Conference on the Psychology of Mathematics Education* (Vol. 2, pp. 281–288). Norwich, UK: IGPME.

Dreyfus, T. (1994). Imagery and reasoning in mathematics and mathematics education. In D. F. Robitalle, D. H. Wheeler, & C. Kieran (Eds.), *Selected Lectures from the 7th International Congress on Mathematical Education* (pp. 107–122). Quebec, Canada: Les Presses de l'Université Laval.

Dubinsky, E., Dautermann, J., Leron, U., & Zazkis, R. (1994). On learning fundamental concepts of group theory. *Educational Studies in Mathematics, 27*, 267–305.

Dubinsky, E., Elterman, F., & Gong, C. (1988). The student's construction of quantification. *For the Learning of Mathematics, 8*(2), 44–51.

Dubinsky, E. & Yiparaki, O. (2001). On student understanding of AE and EA quantification. In E. Dubinsky, A. Schoenfeld, & J. Kaput (Eds.), *Research in Collegiate Mathematics Education IV*. Providence, RI: American Mathematical Society.

Edwards, A. & Alcock, L. (2010). How do undergraduate students navigate their example spaces? In *Proceedings of the 32nd Conference on Research in Undergraduate Mathematics Education*. Raleigh, NC, USA.

Epp, S. (2003). The role of logic in teaching proof. *American Mathematical Monthly, 110*, 886–899.

Epp, S. S. (2004). *Discrete Mathematics with Applications*. Belmont, CA: Thompson-Brooks/Cole.

Even, R. (1993). Subject-matter knowledge and pedagogical content knowledge: Prospective secondary teachers and the function concept. *Journal for Research in Mathematics Education*, *24*, 94–116.

Fischbein, E. (1982). Intuition and proof. *For the Learning of Mathematics*, *3*(2), 9–18.

Giaquinto, M. (2007). *Visual Thinking in Mathematics*. Oxford: Oxford University Press.

Gowers, T. (2002). *Mathematics: A Very Short Introduction*. Oxford: Oxford University Press.

Gray, E. & Tall, D. (1994). Duality, ambiguity and flexibility: A proceptual view of simple arithmetic. *Journal for Research in Maths Education*, *25*, 115–141.

Gueudet, G. (2008). Investigating the secondary–tertiary transition. *Educational Studies in Mathematics*, *67*, 237–254.

Hadamard, J. (1945). *The Psychology of Invention in the Mathematical Field* (1954 edition). New York: Dover Publications.

Hanna, G. (1991). Mathematical proof. In D. O. Tall (Ed.), *Advanced Mathematical Thinking* (pp. 54–61). Dordrecht: Kluwer.

Harel, G. & Sowder, L. (1998). Students' proof schemes: Results from exploratory studies. In A. H. Schoenfeld, J. Kaput, & E. Dubinsky (Eds.), *Research in Collegiate Mathematics III* (pp. 234–282). Providence, RI: American Mathematical Society.

Hazzan, O. & Leron, U. (1996). Students' use and misuse of mathematical theorems: The case of Lagrange's theorem. *For the Learning of Mathematics*, *16*(1), 23–26.

Healy, L. & Hoyles, C. (2000). A study of proof conceptions in algebra. *Journal for Research in Mathematics Education*, *31*(4), 396–428.

Hegarty, M. & Kozhevnikov, M. (1999). Types of visual-spatial representations and mathematical problem solving. *Journal of Educational Psychology*, *91*, 684–689.

Heinze, A. (2010). Mathematicians' individual criteria for accepting theorems and proofs: An empirical approach. In G. Hanna, H. N. Jahnke, & H. Pulte (Eds.), *Explanation and Proof in Mathematics* (pp. 101–111). New York: Springer.

Hemmi, K. (2010). Three styles characterising mathematicians' pedagogical perspectives on proof. *Educational Studies in Mathematics*, *75*, 271–291.

Hernandez-Martinez, P., Black, L., Williams, J., Davis, P., Pampaka, M., & Wake, G. (2008). Mathematics students' aspirations for higher education: Class, ethnicity, gender and interpretative repertoire styles. *Research Papers in Education*, 23, 153–165.

Hersh, R. (1993). Proving is convincing and explaining. *Educational Studies in Mathematics*, 24(4), 389–399.

Higham, N. J. (1998). *Handbook of Writing for the Mathematical Sciences*. Philadelphia, PA: Society for Industrial and Applied Mathematics.

Hoch, M. & Dreyfus, T. (2006). Structure sense versus manipulation skills: An unexpected result. In J. Novotná, H. Moraová, M. Krátká, & N. Stehlíková (Eds.), *Proceedings of the 30th Conference of the International Group for the Psychology of Mathematics Education* (Vol. 3, pp. 305–312). Prague, Czech Republic: PME.

Housman, D. & Porter, M. (2003). Proof schemes and learning strategies of above-average mathematics students. *Educational Studies in Mathematics*, 53, 139–158.

Houston, K. (2009). *How to Think Like a Mathematician*. Cambridge: Cambridge University Press.

Hoyles, C. & Küchemann, D. (2002). Students' understanding of logical implication. *Educational Studies in Mathematics*, 51, 193–223.

Iannone, P., Inglis, M., Mejía-Ramos, J., Simpson, A., & Weber, K. (2011). Does generating examples aid proof production? *Educational Studies in Mathematics*, 77, 1–14.

Inglis, M. (2003). *Mathematicians and the Selection Task*. Unpublished master's thesis, University of Warwick, Warwick, UK.

Inglis, M. & Alcock, L. (2012). Expert and novice approaches to reading mathematical proofs. *Journal for Research in Mathematics Education*, 43, 358–390.

Inglis, M. & Mejia-Ramos, J. P. (2008). How persuaded are you? A typology of responses. *Research in Mathematics Education*, 10, 119–133.

Inglis, M. & Mejia-Ramos, J. P. (2009a). The effect of authority on the persuasiveness of mathematical arguments. *Cognition and Instruction*, 27, 25–50.

Inglis, M. & Mejia-Ramos, J. P. (2009b). On the persuasiveness of visual arguments in mathematics. *Foundations of Science*, 14, 97–110.

Inglis, M., Mejia-Ramos, J. P., & Simpson, A. (2007). Modelling mathematical argumentation: The importance of qualification. *Educational Studies in Mathematics*, 66, 3–21.

Inglis, M., Palipana, A., Trenholm, S., & Ward, J. (2011). Individual differences in students' use of optional learning resources. *Journal of Computer Assisted Learning, 27*, 490–502.

Inglis, M. & Simpson, A. (2008). Conditional inference and advanced mathematical study. *Educational Studies in Mathematics, 67*, 187–204.

Inglis, M. & Simpson, A. (2009). Conditional inference and advanced mathematical study: Further evidence. *Educational Studies in Mathematics, 72*, 185–198.

Jaworski, B. (2002). Sensitivity and challenge in university mathematics tutorial teaching. *Educational Studies in Mathematics, 51*, 71–94.

Johnson-Laird, P. N. & Byrne, R. M. J. (1991). *Deduction*. Hove, UK: Erlbaum.

Kahn, P. E. & Hoyles, C. (1997). The changing undergraduate experience: A case study of single honours mathematics in England and Wales. *Studies in Higher Education, 22*, 349–362.

Kalyuga, S., Ayres, P., Chandler, P., & Sweller, J. (2003). The expertise reversal effect. *Educational Psychologist, 38*, 23–31.

Kember, D. (2004). Interpreting student workload and the factors which shape students' perceptions of their workload. *Studies in Higher Education, 29*, 165–184.

Kember, D. & Kwan, K.-P. (2000). Lecturers' approaches to teaching and their relationship to conceptions of good teaching. *Instructional Science, 28*, 469–490.

Kember, D. & Leung, D. Y. P. (2006). Characterising a teaching and learning environment conducive to making demands on students while not making their workload excessive. *Studies in Higher Education, 29*, 165–184.

Kirshner, D. & Awtry, T. (2004). Visual salience of algebraic transformations. *Journal for Research in Mathematics Education, 35*, 224–257.

Knuth, E. (2002). Secondary school mathematics teachers' conceptions of proof. *Journal for Research in Mathematics Education, 33*, 379–405.

Ko, Y.-Y. & Knuth, E. (2009). Undergraduate mathematics majors' writing performance producing proofs and counterexamples about continuous functions. *Journal of Mathematical Behavior, 28*, 68–77.

Katz, B. P. & Starbird, M. (2012). *Distilling Ideas: An Introduction to Mathematical Thinking*. Mathematical Association of America.

Krantz, S. G. (1997). *Techniques of Problem Solving*. Providence, RI: American Mathematical Society.

Krantz, S. G. (2002). *The Elements of Advanced Mathematics (2nd Edition)*. Boca Raton, FL: Chapman & Hall/CRC.

Kruschke, J. K. (2005). Category learning. In K. Lamberts & R. Goldstone (Eds.), *The Handbook of Cognition* (pp. 183–210). London: Sage.

Krutetskii, V. A. (1976). *The Psychology of Mathematical Abilities in Schoolchildren*. Chicago: University of Chicago Press.

Lakatos, I. (1976). *Proofs and Refutations*. Cambridge: Cambridge University Press.

Lampert, M. (1990). When the problem is not the question and the solution is not the answer: Mathematical knowing and teaching. *American Educational Research Journal*, 27, 29–63.

Larsen, S. (2009). Reinventing the concepts of group and isomorphism: The case of Jessica and Sandra. *Journal of Mathematical Behavior*, 28, 119–137.

Larsen, S. & Zandieh, M. (2008). Proofs and refutations in the undergraduate mathematics classroom. *Educational Studies in Mathematics*, 67, 205–216.

Laurillard, D. (2009). The pedagogical challenges to collaborative technologies. *Computer-Supported Collaborative Learning*, 4, 5–20.

Lawless, C. (2000). Using learning activities in mathematics: Workload and study time. *Studies in Higher Education*, 25, 97–111.

Leikin, R. & Wicki-Landman, G. (2000). On equivalent and non-equivalent definitions: Part 2. *For the Learning of Mathematics*, 20(2), 24–29.

Leinhardt, G., Zaslavsky, O., & Stein, M. K. (1990). Functions, graphs, and graphing: Task, learning, and teaching. *Review of Educational Research*, 60, 1–64.

Leron, U. (1985). A direct approach to indirect proofs. *Educational Studies in Mathematics*, 16, 321–325.

Leron, U., Hazzan, O., & Zazkis, R. (1995). Learning group isomorphism: A crossroads of many concepts. *Educational Studies in Mathematics*, 29, 153–174.

Liebeck, M. (2011). *A Concise Introduction to Pure Mathematics (3rd Edition)*. Boca Raton, FL: CRC Press.

Lin, F.-L. & Yang, K.-L. (2007). The reading comprehension of geometric proofs: The contribution of knowledge and reasoning. *International Journal of Science and Mathematics Education*, 5, 729–754.

Lindlbom-Ylänne, S., Trigwell, K., Nevgi, A., & Ashwin, P. (2006). How approaches to teaching are affected by discipline and teaching context. *Studies in Higher Education*, 31, 285–298.

Lithner, J. (2008). A research framework for creative and imitative reasoning. *Educational Studies in Mathematics, 67*, 255–276.

Lizzio, A., Wilson, K., & Simons, R. (2002). University students' perceptions of the learning environment and academic outcomes: Implications for theory and practice. *Studies in Higher Education, 27*, 27–52.

London Mathematical Society (1995). *Tackling the Mathematics Problem*. London: LMS.

Mann, S. & Robinson, A. (2009). Boredom in the lecture theatre: an investigation into the contributors, moderators and outcomes of boredom among university students. *British Educational Research Journal, 35*, 243–258.

Mariotti, M. A. (2006). Proof and proving in mathematics education. In A. Gutiérrez & P. Boero (Eds.), *Handbook of Research on the Psychology of Mathematics Education: Past, Present and Future* (pp. 173–204). Rotterdam: Sense.

Marton, F. & Säljö, R. (1976). On qualitative differences in learning 1. *British Journal of Educational Psychology, 46*, 4–11.

Mason, J. (2002). *Mathematics Teaching Practice: A Guide for University and College Lecturers*. Chichester: Horwood Publishing.

Mason, J., Burton, L., & Stacey, K. (2010). *Thinking Mathematically (2nd Edition)*. Harlow: Pearson Education.

Mason, J. & Pimm, D. (1984). Generic examples: Seeing the general in the particular. *Educational Studies in Mathematics, 15*, 277–289.

McNamara, D. S., Kintsch, E., Songer, N. B., & Kintsch, W. (1996). Are good texts always better? Interactions of text coherence, background knowledge, and levels of understanding in learning from text. *Cognition and Instruction, 14*, 1–43.

Michener, E. R. (1978). Understanding understanding mathematics. *Cognitive Science, 2*, 361–383.

Monaghan, J. (1991). Problems with the language of limits. *For the Learning of Mathematics, 11*, 20–24.

Moore, R. (1994). Making the transition to formal proof. *Educational Studies in Mathematics, 27*, 249–266.

Moore, S. & Murphy, S. (2005). *How to be a Student: 100 Great Ideas and Practical Habits for Students Everywhere*. Maidenhead: Open University Press.

Movshovitz-Hadar, N. & Hazzan, O. (2004). How to present it? On the rhetoric of an outstanding lecturer. *International Journal of Mathematical Education in Science and Technology, 35*, 813–827.

Muis, K. R. (2004). Personal epistemology and mathematics: A critical review and synthesis of research. *Review of Educational Research, 74*, 317–377.

Nardi, E. (2008). *Amongst Mathematicians: Teaching and Learning Mathematics at University Level*. New York: Springer.

Nelsen, R. B. (1993). *Proofs Without Words: Exercises in Visual Thinking*. Washington, DC: Mathematical Association of America.

Newport, C. (2007). *How to Become a Straight-A Student: The Unconventional Strategies Real College Students Use to Score High While Studying Less*. New York: Three Rivers Press.

Oaksford, M. & Chater, N. (1996). Rational explanation of the selection task. *Psychological Review, 103*, 381–391.

O'Connell, F. (2009). *Work Less, Achieve More: Great Ideas to Get Your Life Back*. London: Headline Publishing Group.

Österholm, M. (2005). Characterizing reading comprehension of mathematical texts. *Educational Studies in Mathematics, 63*, 325–346.

Peled, I. & Zaslavsky, O. (1997). Counter-examples that (only) prove and counter-examples that (also) explain. *Focus on Learning Problems in Mathematics, 19*, 49–61.

Perry, W. G. (1970). *Forms of Intellectual and Ethical Development in the College Years: A Scheme*. New York: Holt, Rinehart and Winston.

Perry, W. G. (1988). Different worlds in the same classroom. In P. Ramsden (Ed.), *Improving Learning: New Perspectives* (pp. 145–161). London: Kogan Page.

Pinto, M. & Tall, D. O. (2002). Building formal mathematics on visual imagery: A case study and a theory. *For the Learning of Mathematics, 22*, 2–10.

Poincaré, H. (1905). *Science and Hypothesis*. London: Walter Scott Publishing.

Pólya, G. (1957). *How to Solve It: A New Aspect of Mathematical Method (2nd Edition)*. Princeton, NJ: Princeton University Press.

Presmeg, N. (2006). Research on visualization in learning and teaching mathematics. In A. Gutiérrez & P. Boero (Eds.), *Handbook of Research on the Psychology of Mathematics Education: Past, Present and Future* (pp. 205–235). Rotterdam: Sense.

Raman, M. (2003). Key ideas: What are they and how can they help us understand how people view proof? *Educational Studies in Mathematics*, *52*, 319–325.

Raman, M. (2004). Epistemological messages conveyed by three high-school and college mathematics textbooks. *Journal of Mathematical Behavior*, *23*, 389–404.

Ramsden, P. (2003). *Learning to Teach in Higher Education*. Abingdon: RoutledgeFalmer.

Rav, Y. (1999). Why do we prove theorems? *Philosophia Mathematica*, *7*, 5–41.

Recio, A. & Godino, J. (2001). Institutional and personal meanings of mathematical proof. *Educational Studies in Mathematics*, *48*, 83–99.

Reid, D. A. & Knipping, C. (2010). *Proof in Mathematics Education: Research, Learning and Teaching*. Rotterdam: Sense Publishers.

Rips, L. (1994). *The Psychology of Proof: Deductive Reasoning in Human Thinking*. Cambridge, MA: MIT Press.

Rowland, T. (2002). Generic proofs in number theory. In S. R. Campbell & R. Zazkis (Eds.), *Learning and Teaching Number Theory: Research in Cognition and Instruction* (pp. 157–184). Westport, CT: Ablex Publishing Corp.

Ryan, J. & Williams, J. (2007). *Children's Mathematics 4–15: Learning from Errors and Misconceptions*. Maidenhead: Open University Press.

Sangwin, C. J. (2003). New opportunities for encouraging higher level mathematical learning by creative use of emerging computer aided assessment. *International Journal of Mathematical Education in Science and Technology*, *34*, 813–829.

Schoenfeld, A. H. (1985). *Mathematical Problem Solving*. San Diego: Academic Press.

Schoenfeld, A. H. (1992). Learning to think mathematically: Problem solving, metacognition and sense making in mathematics. In D. Grouws (Ed.), *Handbook of Research on Mathematics Teaching and Learning* (pp. 334–370). New York: Macmillan.

Seely, J. (2004). *Oxford A–Z of Grammar & Punctuation*. Oxford: Oxford University Press.

Segal, J. (2000). Learning about mathematical proof: Conviction and validity. *Journal of Mathematical Behavior*, *18*(2), 191–210.

Selden, A. & Selden, J. (1999). *The Role of Logic in the Validation of Mathematical Proofs* (Tech. Rep.). Cookeville, TN, USA: Tennessee Technological University.

Selden, A. & Selden, J. (2003). Validations of proofs considered as texts: can undergraduates tell whether an argument proves a theorem? *Journal for Research in Mathematics Education*, *34*(1), 4–36.

Selden, J. & Selden, A. (1995). Unpacking the logic of mathematical statements. *Educational Studies in Mathematics*, *29*, 123–151.

Sfard, A. (1991). On the dual nature of mathematical conceptions: Reflections on processes and objects as different sides of the same coin. *Educational Studies in Mathematics*, *22*, 1–36.

Skemp, R. R. (1976). Relational understanding and instrumental understanding. *Mathematics Teaching*, *77*, 20–26.

Smith, G., Wood, L., Coupland, M., Stephenson, B., Crawford, K., & Ball, G. (1996). Constructing mathematical examinations to assess a range of knowledge and skills. *International Journal of Mathematical Education in Science and Technology*, *27*, 65–77.

Solomon, Y., Croft, T., & Lawson, D. (2010). Safety in numbers: Mathematics support centres and their derivatives as social learning spaces. *Studies in Higher Education*, *35*, 421–431.

Solow, D. (2005). *How to Read and Do Proofs*. Hoboken, NJ: John Wiley.

Speer, N. M., Smith, J. P. III, & Horvath, A. (2010). Collegiate mathematics teaching: An unexamined practice. *Journal of Mathematical Behavior*, *29*, 99–114.

Sperber, D. & Wilson, D. (1986). *Relevance: Communication and Cognition*. London: Blackwell.

Stanovich, K. E. (1999). *Who is Rational? Studies of Individual Differences in Reasoning*. Mahwah, NJ: Lawrence Erlbaum.

Stewart, I. (1995). *Concepts of Modern Mathematics*. New York: Dover Publications.

Stewart, I. (2006). *Letters to a Young Mathematician*. New York: Basic Books.

Stewart, I. N. & Tall, D. O. (1977). *The Foundations of Mathematics*. Oxford: Oxford University Press.

Strunk Jr., W. & White, E. B. (1999). *The Elements of Style (4th Edition)*. Longman.

Stylianides, A. J. (2007). Proof and proving in school mathematics. *Journal for Research in Mathematics Education*, *38*, 289–321.

Stylianides, A. J. & Stylianides, G. J. (2009). Proof constructions and evaluations. *Educational Studies in Mathematics*, *72*, 237–253.

Stylianou, D. A. & Silver, E. A. (2004). The role of visual representations in advanced mathematical problem solving: An examination of expert–novice similarities and differences. *Mathematical Thinking and Learning*, *6*, 353–387.

Swinyard, C. (2011). Reinventing the formal definition of limit: The case of Amy and Mike. *Journal of Mathematical Behavior*, *30*, 93–114.

Tall, D. O. (1989). The nature of mathematical proof. *Mathematics Teaching*, *127*, 28–32.

Tall, D. O. (1995). Cognitive development, representations and proof. In *Proceedings of Justifying and Proving in School Mathematics* (pp. 27–38). London: Institute of Education.

Tall, D. O. & Vinner, S. (1981). Concept image and concept definition in mathematics with particular reference to limits and continuity. *Educational Studies in Mathematics*, *12*, 151–169.

Thurston, W. P. (1994). On proof and progress in mathematics. *Bulletin of the American Mathematical Society*, *30*, 161–177.

Toulmin, S. (1958). *The Uses of Argument*. Cambridge: Cambridge University Press.

Trask, R. L. (1997). *The Penguin Guide to Punctuation*. London: Penguin.

Trask, R. L. (2002). *Mind the Gaffe: The Penguin Guide to Common Errors in English*. London: Penguin Books.

Trigwell, K. & Prosser, M. (2004). Development and use of the approaches to teaching inventory. *Educational Psychology Review*, *16*, 409–424.

Tsamir, P., Tirosh, D., & Levenson, E. (2008). Intuitive nonexamples: The case of triangles. *Educational Studies in Mathematics*, *49*, 81–95.

Usiskin, Z., Peressini, A., Marchisotto, E. A., & Stanley, D. (2003). *Mathematics for High School Teachers: An Advanced Perspective*. Upper Saddle River, NJ: Prentice Hall.

Vamvakoussi, X., Christou, K. P., Mertens, L., & Van Dooren, W. (2011). What fills the gap between discrete and dense? Greek and Flemish students' understanding of density. *Learning and Instruction*, *21*, 676–685.

Van Dooren, W., de Bock, D., Weyers, D., & Verschaffel, L. (2004). The predictive power of intuitive rules: A critical analysis of 'more A–more B' and 'same A–same B'. *Educational Studies in Mathematics*, *56*, 179–207.

Van Dormolen, J. & Zaslavsky, O. (2003). The many facets of a definition: The case of periodicity. *Journal of Mathematical Behavior*, *22*, 91–106.

Velleman, D. J. (2004). *How to Prove It: A Structured Approach.* Cambridge: Cambridge University Press.

Vermetten, Y. J., Lodewijks, H. G., & Vermunt, J. D. (1999). Consistency and variability of learning strategies in different university courses. *Higher Education, 37,* 1–21.

Vermunt, J. D. & Verloop, N. (1999). Congruence and friction between learning and teaching. *Learning and Instruction, 9,* 257–280.

Vinner, S. (1991). The role of definitions in teaching and learning. In D. O. Tall (Ed.), *Advanced Mathematical Thinking* (pp. 65–81). Dordrecht: Kluwer.

Vinner, S. & Dreyfus, T. (1989). Images and definitions for the concept of function. *Journal for Research in Mathematics Education, 20,* 356–366.

Vivaldi, F. (2011). *Mathematical Writing: An Undergraduate Course.* Online at http://www.maths.qmul.ac.uk/~fv/books/mw/mwbook.pdf.

Weber, K. (2001). Student difficulty in constructing proofs: the need for strategic knowledge. *Educational Studies in Mathematics, 48,* 101–119.

Weber, K. (2004). Traditional instruction in advanced mathematics courses: A case study of one professor's lectures and proofs in an introductory real analysis course. *Journal of Mathematical Behavior, 23,* 1151–33.

Weber, K. (2005). On logical thinking in mathematics classrooms. *For the Learning of Mathematics, 25*(3), 30–31.

Weber, K. (2008). How mathematicians determine if an argument is a valid proof. *Journal for Research in Mathematics Education, 39,* 431–459.

Weber, K. (2009). How syntactic reasoners can develop understanding, evaluate conjectures, and generate examples in advanced mathematics. *Journal of Mathematical Behavior, 28,* 200–208.

Weber, K. (2010a). Mathematics majors' perceptions of conviction, validity and proof. *Mathematical Thinking and Learning, 12,* 306–336.

Weber, K. (2010b). Proofs that develop insight. *For the Learning of Mathematics, 30,* 32–36.

Weber, K. & Alcock, L. (2004). Semantic and syntactic proof productions. *Educational Studies in Mathematics, 56,* 209–234.

Weber, K. & Alcock, L. (2005). Using warranted implications to understand and validate proofs. *For the Learning of Mathematics, 25*(1), 34–38.

Weber, K. & Alcock, L. (2009). Proof in advanced mathematics classes: Semantic and syntactic reasoning in the representation system of proof. In

D. A. Stylianou, M. L. Blanton, & E. Knuth (Eds.), *Teaching and Learning Proof Across the Grades: A K-16 Perspective* (pp. 323–338). New York: Routledge.

Weber, K. & Mejia-Ramos, J. P. (2009). An alternative framework to evaluate proof productions: A reply to Alcock and Inglis. *Journal of Mathematical Behavior, 28,* 212–216.

Weber, K. & Mejia-Ramos, J.-P. (2011). Why and how mathematicians read proofs: An exploratory study. *Educational Studies in Mathematics, 76,* 329–344.

Weinberg, A. & Wiesner, E. (2011). Understanding mathematics textbooks through reader-oriented theory. *Educational Studies in Mathematics, 76,* 49–63.

Wicki-Landman, G. & Leikin, R. (2000). On equivalent and non-equivalent definitions: Part 1. *For the Learning of Mathematics, 20*(1), 17–21.

Williams, C. G. (1998). Using concept maps to assess conceptual knowledge of function. *Journal for Research in Mathematics Education, 29,* 414–421.

Yackel, E., Rasmussen, C., & King, K. (2000). Social and sociomathematical norms in an advanced undergraduate mathematics course. *Journal of Mathematical Behavior, 19,* 275–287.

Yang, K.-L. & Lin, F.-L. (2008). A model of reading comprehension of geometry proof. *Educational Studies in Mathematics, 67,* 59–76.

Yusof, Y. B. M. & Tall, D. O. (1999). Changing attitudes to university mathematics through problem solving. *Educational Studies in Mathematics, 37,* 67–82.

Zandieh, M. & Rasmussen, C. (2010). Defining as a mathematical activity: A framework for characterizing progress from informal to more formal ways of reasoning. *Journal of Mathematical Behavior, 29,* 57–75.

Zaslavsky, O. & Shir, K. (2005). Students' conceptions of a mathematical definition. *Journal for Research in Mathematics Education, 36,* 317–346.

Zazkis, R. & Chernoff, E. J. (2008). What makes a counterexample exemplary? *Educational Studies in Mathematics, 68,* 195–208.

INDEX

abbreviation, 73, 136, 139
ability, 231
abroad, 192
Abstract Algebra, 34, 119
academics, 242
accommodation office, 197
accuracy, 137
adapting proofs, 123, 124
addition, 31, 47
additional needs, 197
additive identity, 118
administration, 242
adviser
  project, 192
advisers, xvi
all, 70
always, 61
ambiguity, 71, 116, 153, 165
Analysis, 48, 94, 119, 138, 142
and so on, 95
angle, 144
anonymity, 179
applicability of a method, 122
applied mathematics, 17, 243
approximation, 110
arbitrary, 81, 83
argument
  backward, 159
  conventions in, 155
  formal, 91
  informal, 108
  layout, 155, 165
arithmetic, 20
  new kinds of, 34
arrows, 163

as required, 83
assignments, xvi, 203, 205
assuming what trying to prove, 156, 159
assumption, 40
  accidental, 13
  temporary, 108
assurance, 121
axioms, 37, 94

backward argument, 159
base case, 115
basis, 25
best student, xii, 230, 231
bigger than, 91
binary operation, 31, 32, 46, 47
brackets, 66, 163, 164
break, 100
brevity, 108, 118, 137

calculations, 4, 12, 100, 123
capturing meaning, 55
car, 69
Cardiff, 69
careers, 154, 193, 195, 196, 246
Cartesian graph, 44
cases, 105, 107
catching up, 225, 233
caution, 97, 124
certainty, 93
chain of equalities/inequalities, 107, 115
chain rule, 90
challenge, xiii, 236, 237
change signs, 15
check/question-mark/cross, 140
choosing examples, 51